A History Of Jet Propulsion, Including Rockets

A History Of Jet Propulsion, Including Rockets

Raymond Friedman

Library of Congress Control Number: 2010903902
ISBN: Hardcover 978-1-4500-6590-0
Softcover 978-1-4500-6589-4
Ebook 978-1-4500-6591-7

This book was printed in the United States of America.

To order additional copies of this book, contact:
Xlibris Corporation
1-888-795-4274
www.Xlibris.com
Orders@Xlibris.com
78360

ACKNOWLEDGEMENTS

I have been lucky in being able to associate during my career with a number of key figures in development of jet propulsion and rocketry: Joe Hirschfelder, my thesis advisor at Wisconsin, who pioneered in developing computer analysis of combustion; Stewart Way, my supervisor at Westinghouse Research, who was an inventor of the ramjet; Arch Scurlock, president of Atlantic Research, who was an early supporter of the U.S.Navy's use of composite solid propellants in submarines launched rockets; John Fenn, who led many of the Office of Naval Research's combustion research programs; and Theodore von Karman, Chief Scientist of the U.S. Air Force.

The encouragement of my wife, Myra, was important to me in writing this book.

CONTENTS

1

INTRODUCTION

In the history of technology development, many advances occurred by trial and error, or by accidental discoveries. However, quite a number of advances, including that of jet propulsion, resulted from new scientific knowledge.

As an example of progress by trial and error, consider metallurgy; the discovery of brass, bronze, iron, and steel. These metals were discovered long before there was any true knowledge of the chemical elements or the chemistry of refining ores. Another example was the accidental discovery of radium by the Curies. Perhaps the earliest important example of an accidental discovery was how to control fire, probably from observation of fire started by lightning. The discovery of agriculture (growing things by spreading seeds around) occurred long before the evolution of the science of biology.

On the other hand, many discoveries resulted from scientific advances. For example, James Clerk Maxwell in 1864 developed the theory of electromagnetism, which predicted that an electromagnetic effect would produce a "field" which would spread through space with the speed of light. This unexpected prediction led directly to wireless telegraphy, followed by radio and television. As another example, Albert Einstein in 1905 put forward the concept that mass could change into energy. This explained where the sun's energy came from, and, more practically, led to nuclear power plants. Alfred Wegener's discovery of continental drift provided a key to the prediction of earthquakes.

What about jet propulsion? This is based on Newton's laws of motion. Newton's work, in turn, followed Galileo's demonstration that the acceleration of falling bodies, as well as the swinging of a pendulum, followed mathematical laws. At the same time the astronomer Kepler was able to provide a precise explanation of the motions of the planets by postulating movements in ellipses, with the sun at a focus, and the time of each orbit being proportional to the three-halves power of the mean distance from the sun. Isaac Newton, having first discovered calculus, built on Galileo's and Kepler's discoveries to establish the laws of motion and gravitation. From these laws, it was clear how a rearward jet provided forward propulsion. Furthermore, rocket speeds necessary to go into orbit around the earth or to escape the earth could be easily calculated.

The development of fixed-wing airplanes was based on the scientist Jacob Bernoulli's discovery that air flowing over and under a concave-downward wing would produce a pressure difference providing lift.

Before jet propulsion was introduced into aviation, airplane speeds were limited to around 350 miles per hour, because, at higher speeds than this, shock waves would form at the propeller tips, causing a major decrease in propeller efficiency. (The speed of sound in air is about 700 miles per hour, and this speed is reached by the combined motion of the plane and the propeller tip.) Jet planes (no propeller) first flew in 1939.

Before powerful liquid-propellant rockets and rocket staging were developed, it was not possible for a rocket to escape the earth's gravitational field. The staging concept was developed by the mathematician Tsiolkovsky early in the twentieth century, and the first practical liquid-propellant engines were developed in the 1930's.

Accordingly, the decade of the 1930's marked a turning point in both aircraft propulsion and rocketry. By the 1960's supersonic aircraft were flying routinely. Rockets were traveling to the moon.

Both jet engines and rocket engines work by the principle of the three laws of motion, first clarified by Isaac Newton in the seventeenth century.

This book will describe the progress, step by step, in developing these two modes of propulsion.

Mankind, observing the flight of birds, has always been interested in flying. Leonardo da Vinci, in the sixteenth century, sketched the design of a flying machine. However this did not lead to a successful flight. The first actual flight was made by P. de Rozier and M. D'Arlandes on November 23, 1783 in Paris, in a hot-air balloon designed by the Montgolfier brothers. See Figure 1. Many balloon flights followed, always at the mercy of the winds.

FIGURE 1. MONTGOLFIER'S HOT AIR BALLOON

Another technique for flying was introduced in 1853, a manned glider designed by Sir George Cayley, an Englishman. Cayley's design was based on the discovery by the Swiss scientist Jacob Bernoulli that a fixed wing can support weight if it is curved and concave downward. As air flows over the wing, the portion flowing over the top will increase in speed and decrease in pressure, while the portion flowing beneath the wing will decrease in speed and increase in pressure. This pressure difference provides "lift".

Many successive glider flights were made, notably by Otto Lilienthal. He made hundreds of such flights, until killed in a crash.

The next significant development was the internal combustion engine, first built by Etienne Lenoir in France in 1860. By 1885, such engines were incorporated into automobiles. Internal combustion engines could be light enough to power an airplane propeller. Within a few years, Ferdinand von Zeppelin in Germany added a gasoline engine driving a propeller to a balloon, thereby producing the first controllable aircraft. However, such a craft was slow and bulky, and subject to damage by storms.

At the turn of the century, numerous inventors in many countries were trying to combine a glider with a gasoline engine. The first success was scored by the Wright brothers in 1903, at Kitty Hawk, North Carolina. See Figure 2 and Appendix C, for details of how the Wright brothers solved the problem by the use of control surfaces. From then on, there was rapid improvement in airplane design. Glenn Curtiss, called the father of the U.S. aviation industry, started the Curtiss Manufacturing Company to manufacture airplane engines in 1905. He worked with the U.S. Navy to build the first plane to take off from a ship, in 1910. In 1911 he built the first seaplane, with floats as well as wheels.

FIGURE 2. PICTURE OF WRIGHT BROTHERS' PLANE

Airplanes were widely used in World War I, chiefly for reconnaissance. Scheduled transcontinental airmail service was commenced in 1921.

In 1924, two U.S. Army airplanes made a round-the-world trip, but it took 35 days. In 1931, Wiley Post and Harold Gatty's round-the-world trip took only four days. By the time of World War II, larger and faster planes were developed, and widely used in the war for bombing of distant targets. Planes from aircraft carriers were able to sink battleships.

At this point in history, it appeared that plane design had reached a limit of speed, around 350 miles per hour, because at higher speeds the propeller tips were moving through air faster than the speed of sound. This means that shock waves were generated, with accompanying severe vibrations and major increase in air resistance.

The velocity of sound in air six miles above sea level is 693 miles per hour, and the temperature is about 40 degrees below zero (Fahrenheit). In air, the molecules are moving rapidly and randomly, frequently colliding with one another. Their average velocity at 40 degrees below zero is 685 miles per hour. It is immediately obvious why air cannot flow smoothly around an object when the relative speed is greater than the speed of sound.

Accordingly, something other than a propeller is needed to propel an airplane through the sound barrier. That "something" turns out to be a jet engine, invented in the 1930's. Let us take a minute to discuss why high speed in an airplane is desirable.

The military advantages of a high-speed aircraft are obvious. A fighter plane can more easily intercept an enemy, and a fast-moving bomber is harder to be intercepted. The chief benefit to civilian aircraft is that they can carry their passengers to their destinations more quickly. There is an additional benefit; the faster a plane can fly, the higher it can fly, because the lift of the wing is greater at higher speeds. The plane at high altitude is moving through more rarefied air, which offers less resistance, leading to a fuel saving. The plane may be able to fly above turbulent weather. If going east, it may be able to take advantage of the jet stream, moving at more than one hundred miles per hour. If going west, it may be able to avoid the jet stream. A high-flying military plane is harder to attack with an anti-aircraft missile.

There are three ways that jet propulsion may be used; with turbojets, ramjets, and rockets. A turbojet engine compresses the inlet air, heats the compressed air by combustion, extracts energy from it with a turbine, and then exhausts it through a nozzle. A ramjet, which can only operate supersonically, slows down the entering air with shock waves, then heats the air by combustion, and finally exhausts it through a nozzle. The air may pass through the combustion chamber either subsonically or supersonically. A turbojet engine has rotating moving parts, a compressor and a turbine, while a ramjet engine has no moving parts.

A rocket makes no use of the outside air. Both the fuel and the oxidant are on board, and are burned at high pressure in a combustion chamber. The hot gases are then permitted to exit through a nozzle. A liquid-propellant rocket has fuel and oxidant pumps, while a solid-propellant rocket has no moving parts. A rocket can operate better in a vacuum than in the atmosphere.

Rockets have been around for a long time. A thousand years ago, the Chinese had devised rockets in order to project fireworks into the air. Much later, in the seventeenth and eighteenth centuries, various European countries were using rockets as a military weapon. You may recall "the rockets' red glare, and bombs bursting in air, gave proof through the night that our flag was still there." These were called Congreve rockets, named after a British officer named William Congreve.

However, such solid-propellant rockets did not amount to a very significant weapon, because they were unguided and could only carry a small amount of explosive as a payload. In World War II, the bazooka was used as an important anti-tank rocket, with much success. The great advantage of the bazooka compared with a cannon was that it could be carried and fired by a single soldier. It consisted of a rocket in a tube about three inches in diameter and five feet long, which could be shoulder-fired. The warhead contained enough explosive to destroy a tank. One interesting feature was the means of guiding the rocket, once launched, to the target. This could be accomplished by a wire trailing from the rocket to the launch point. Signals could be sent through this wire to correct the trajectory, assuring a direct hit. Nowadays, guidance might be by radio or laser. See Figure 3.

FIGURE 3. PICTURE OF A BAZOOKA

Other rocket weapons were used in the war, which were larger and capable of carrying a larger payload than the Congreve rockets. One example was the Katusha rockets used by the Russians against the Germans.

An interesting use of rockets during the World War II period was to attach them under the wings of aircraft to provide extra thrust during takeoff, thus avoiding the need of a long runway, as shown in Figure 4.

FIGURE 4. PICTURE OF ROCKET-ASSISTED TAKE-OFF OF A B47 PLANE

In 1944, the Germans commenced the use of the V2 rocket, which used liquid propellants and had a far greater range than any rocket built previously. It could be launched from the European continent and hit London. Its speed was so great that it could not be intercepted by any airplane. However there was a problem with guidance. The Germans could aim it well enough to hit somewhere in London, but it could not be guided to hit a specific target. See Figure 5.

FIGURE 5. PICTURE OF A V2 ROCKET

In 1945, the nuclear bomb came into existence. While the first nuclear bombs were dropped from aircraft, which in theory could have been intercepted, it became obvious that a long-range rocket carrying a nuclear bomb would be an extremely effective weapon, being interceptable only with great difficulty, if at all. This caused a tremendous acceleration in the development of rockets that had longer range and better guidance than the German V2 rocket. The United States and the U.S.S.R. were the leading competitors in this development, and by 1957 both countries had produced rockets capable of hitting a target anywhere on earth, and even capable of launching a payload into orbit around the earth. Thus, the age of space travel was launched.

The first manned orbit was accomplished in 1959, and the first manned trip to the moon and return was accomplished in 1969. Then, unmanned rocket probes were launched to Mars, Venus, and other planets, and plans were made for a manned round trip to Mars.

This book provides the background for these developments.

2

HOW DOES A JET ENGINE PROPEL AN AIRCRAFT?

A propeller-driven airplane is limited to speeds around 350 miles per hour, because of shock waves forming at the propeller tips at greater speeds. How, then, can higher airplane speeds, permitting flight at higher altitudes through thinner air, be accomplished?

This required a revolutionary invention, the turbojet engine, first patented by Frank Whittle in 1930.

How can a jet engine provide thrust to a plane without a propeller? Think of a tube, with air entering the front end with a certain velocity (the flight velocity) and leaving the back end at a higher velocity. According to Isaac Newton's laws of motion, this results in a thrust force propelling the plane forward. The way this follows from Newton's laws is explained in Appendix A.

How is the air able to exit at a higher velocity than its entry velocity? There are two ways of accomplishing this, involving a turbojet or a ramjet. The first engines developed were turbojets, which will be described now, ramjets later.

Inside this tube representing a turbojet engine are four elements; a compressor, a combustion chamber, a turbine, and an exhaust nozzle. (In some cases, an afterburner may be added between the turbine and the exhaust nozzle.) The compressor and the turbine are mounted on the same shaft, so that the power generated by the turbine may be transmitted directly to the compressor. The compressor has several stages, and is generally an axial-flow compressor rather than a centrifugal compressor, because it has smaller frontal area.

The air entering at the front is compressed to five or ten times the external pressure, and then a fuel, usually a hydrocarbon, is sprayed into a combustion chamber and combustion takes place. The fuel and air are not in the "stoichiometric" proportions, corresponding to complete consumption of the fuel and the oxygen, because the resulting temperature would be too high for the turbine to tolerate. An excess of air is provided beyond what is needed to burn the fuel. The combustion products then flow past several stages of turbine blades, producing the power to operate the compressor. The pressure downstream of the turbine is lower than the pressure in the combustion chamber, but higher than the external pressure. The combustion products are then exhausted through a nozzle. The set-up is shown in Figure 6.

FIGURE 6. SKETCH OFA TURBOJET ENGINE

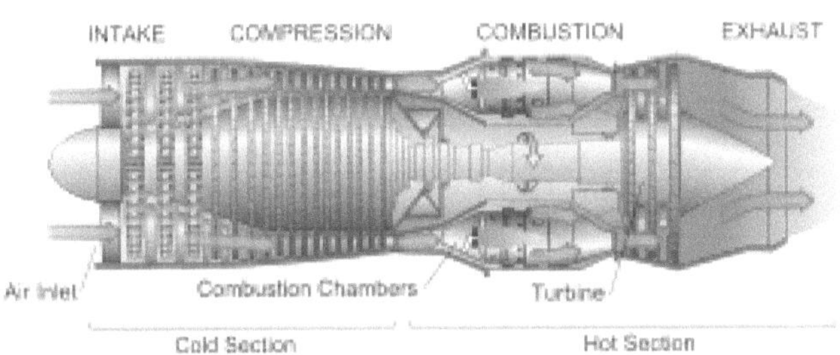

The key to the operation is the expansion of the gases as a result of the heat added by combustion. That is, the volume of the gases passing through the turbine is much greater than the volume of air entering the combustion chamber. As a result, the gases are exiting from the turbine while the pressure is still substantially higher than the external pressure, although lower than in the combustion chamber. The exit nozzle expands these gases to the external pressure, causing the gases to accelerate, and they leave the engine at considerably higher velocity than the velocity at which they entered the engine (the flight velocity).

One modification used for military applications is to burn additional fuel in an afterburner, downstream of the turbine and upstream of the nozzle. This provides an additional burst of power, but is a very inefficient way to use fuel.Such a turbojet engine is capable to propel a plane to high subsonic velocities or even supersonic velocities greater than Mach 2 (twice the speed of sound).

For subsonic flight, a modification of the turbojet called the fanjet is often used. In such an engine, a minor part of the entering air is drawn into the compressor, combustion chamber, and turbine, while the majority of the air is compressed moderately by a fan and then bypassed by the compressor, combustion chamber, and turbine, and finally mixed with the turbine outlet gases and exhausted through the exit nozzle. This results in improved fuel economy.

In order to fly at high supersonic speeds, the ramjet is available. In one type of ramjet, the entering air, at supersonic speed, is slowed down by shock waves to a subsonic velocity. Then, fuel is burned, and the expanded gases are allowed to achieve supersonic velocity with a nozzle. The exit velocity is always higher than the flight velocity, thus providing the thrust force. See Figure 7.

FIGURE 7. SKETCH OF A RAMJET ENGINE

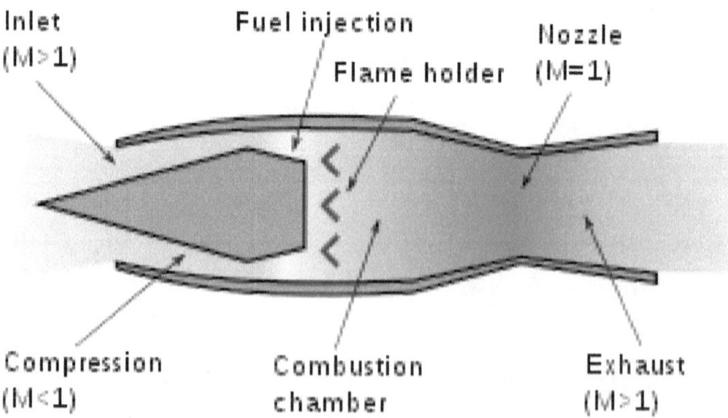

A ramjet engine is not self-starting from rest, but must be accelerated to at least Mach 2 by a turbojet engine or a rocket. It can then operate up to speeds of at least Mach 6.

For even higher flight speeds, another type of ramjet is available, a supersonic combustion ramjet (often called a SCRAMJET). In this version, the entry air is slowed down but remains supersonic throughout the combustion region. This poses the problem of how to accomplish stable combustion in a supersonic flow, since little time is available. This problem has been solved by using hydrogen as a fuel. Hydrogen is easy to ignite and burns rapidly, Theoretically, a SCRAMJET could fly as fast as Mach 17.

A problem of supersonic flight is that of aerodynamic heating of the outside of the vehicle. As the vehicle travels through the air, some of the air is brought to rest on the front surface of the vehicle, and rises in temperature as it does so. It reaches what is called a stagnation temperature, values of which are shown in Figure 8, as a function of Mach number. The external air temperature is assumed to be minus forty degrees Fahrenheit, in accordance with conditions at six miles altitude. The Mach number is the ratio of the air velocity to the local velocity of sound. Hence, Mach numbers greater than one denote supersonic flow.

FIGURE 8. PLOT OF STAGNATION TEMPERATURE VERSUS MACH NUMBER
(assumes air is initially at minus 40 degrees Fahrenheit)

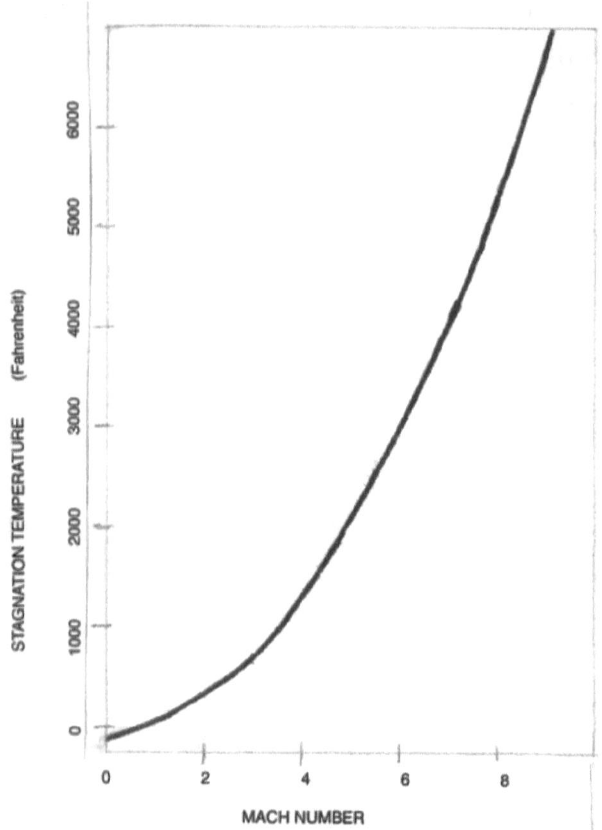

Figure 8 shows that the stagnation temperature at Mach 5 is over 2000 degrees Fahrenheit. Accordingly, at this and higher Mach numbers, provision has to be made to cool the vehicle skin. This may be accomplished by circulating the fuel, possibly liquid hydrogen, near critical regions before injecting it into the combustion chamber. Another possible fuel is liquid methane. Any other hydrocarbon fuel would decompose while being used as coolant.

3

HOW DOES A ROCKET PROPEL ITSELF?

Consider Figure 9 below, which shows, in highly simplified form, the essence of how a rocket works. The pressure inside the combustion chamber must be substantially higher than the outside pressure. This internal high pressure is generated by combustion of either a liquid or a solid fuel (not shown). The hot, high-pressure gas so generated is allowed to expand through a nozzle, as shown. (Combustion is not absolutely necessary; a rocket could be propelled by a stored high-pressure gas, although less effectively.) The gas issues from the nozzle at high velocity.

FIGURE 9. SKETCH OF A LIQUID PROPELLANT ROCKET

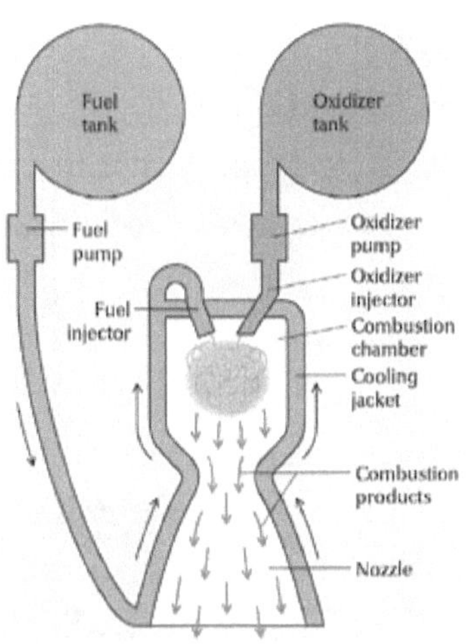

The propulsion force is a direct consequence of the laws of motion first understood by Sir Isaac Newton in the latter part of the seventeenth century. A detailed explanation is provided in Appendix A. This force, called the thrust of the rocket, is proportional to the product of the velocity and the rate of mass flow of the gases exiting the nozzle. (This is strictly true only if the pressure of the gases as they exit the nozzle is identical with the external pressure.)

Notice that the mass of the rocket is becoming smaller as long as the propellant is being expelled, because the mass of the rocket includes the mass of the propellant not yet consumed at any instant. The formula for calculating the velocity imparted to the rocket, derived by the use of calculus, is presented in Appendix B.

The principle by which this propulsion occurs may perhaps be understood by some examples. First, consider a person wearing ice skates, standing on smooth ice. He is holding a bag of sand. He throws the bag of sand forward. Simultaneously, he will be forced to slide backward. The product of the mass of the bag and its velocity will be exactly equal to the product of the mass of the person and his backward velocity according to Newton's laws, as long as the friction of his skates on the ice is negligible.

Second, imagine a bullet being fired from a gun. The gun will recoil. The impulse causing this recoil is equal to the product of the mass of the bullet and its velocity, according to Newton.

Third, imagine a large fire hose discharging 200 gallons per minute. The backward force on the nozzle will be around 100 pounds, and it usually takes two firemen to hold the nozzle.

Using these principles, we can calculate how fast the rocket is accelerating, but we must also take into account the air resistance, unless the rocket is in outer space, and the effect of gravitation, as long as the rocket is in the earth's gravitational field. Thus, from basic principles of physics, we can calculate how the rocket behaves. The main unknown is the velocity with which the exhaust gas leaves the rocket nozzle.

Notice that a rocket can operate better in a vacuum than in the atmosphere, for two reasons. There is no air resistance, and the exhaust gas can leave the rocket at a higher velocity when flowing into a vacuum.

The view expressed by skeptics long ago , including the New York Times, that a rocket can not operate in a vacuum because there is nothing to push against is fallacious. After successful NASA flights, the New York Times published an apology.

A nozzle is an essential part of a rocket. See Figure 10.

FIGURE 10. A CONVERGENT-DIVERGENT NOZZLE

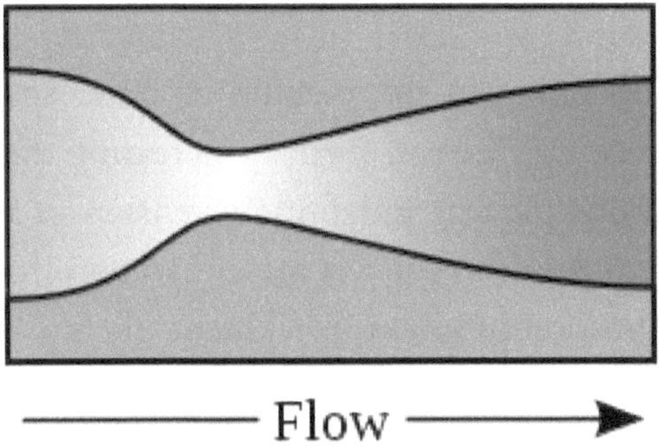

The convergent-divergent nozzle shown in the sketch is known as a de Laval nozzle. It was invented by a Swedish engineer, Gustaf de Laval, in the nineteenth century. He wasn't concerned with rockets; his interest was related to steam turbines. The figure shows a common design of the divergent portion of the nozzle, a so-called bell shape. An alternate design of the divergent portion is a cone shape. While the cone shape is simpler to manufacture, the bell shape has the advantage that it extracts a few percent more thrust from the exiting gas by producing an exit velocity with a direction more precisely aligned with the longitudinal axis of the rocket.

This type of nozzle works as follows. The inlet gas, starting at a high pressure, expands through the convergent section of the nozzle, accelerating as it expands. The gas reaches sonic velocity at the throat of the nozzle, at which point the pressure is about half the initial pressure. It continues to expand and accelerate in the divergent section of the nozzle, moving at increasing supersonic velocities, with decreasing pressure. For a rocket operating in the earth's atmosphere, it is usual to design the rocket so that the exhaust pressure has decreased to match the external atmospheric pressure at the end of the nozzle. A formula for calculating the exhaust velocity under this condition is given in Appendix B. For a rocket, we want this exhaust velocity to be high. The formula tells us that the exhaust velocity will be high if the temperature and pressure in the combustion chamber and the velocity of sound in the exhaust gas are high. A high velocity of sound in the exhaust gas is achieved by adjusting the propellant composition so that the mean molecular weight is as low as possible. The higher the proportion of hydrogen in the exhaust, the lower the mean molecular weight.

For a rocket operating in the vacuum of outer space, the divergent portion of the nozzle may extend further, increasing the exhaust velocity. Weight considerations prevent indefinite extension of the nozzle. Firing into a vacuum, it is possible to convert more than two-thirds of the thermal energy of the combustion products into kinetic energy.

The pressure in the combustion chamber is limited by the strength of the walls, which must be light. The temperature in the combustion chamber as well as the velocity of sound of the gases is determined by the chemical composition of the propellant (or combination of two propellants, fuel and oxidizer). Discussion of the various possible propellants, and the exhaust velocity provided by each propellant and set of operating conditions, is provided later.

The high temperature produces a problem in the design of a rocket, namely the deleterious effect on the combustion chamber, and most critically at the rocket nozzle. There are three approaches to this problem: (1) accept a short burning period, a few tens of seconds; (2) for liquid propellants, use the fuel or oxidizer to cool critical locations, by circulating it through cooling coils on the outside of critical areas, and then inject the warmed, partially vaporized liquid into the combustion chamber; and (3) for solid propellants, since no liquid coolant is available, the approach is to line the combustion chamber and especially the throat of the nozzle with special materials which erode slowly enough to last the necessary time (perhaps 100 seconds). The combustion chamber of a solid-propellant rocket is generally made with continuous filaments of strong fibers such as carbon fibers (best) or Kevlar fibers (next best) held together with a plastic such as an epoxy resin. The best material for the throat of the nozzle is carbon fibers in a carbon matrix.

Another problem with a rocket is how to control the magnitude of the thrust, and how to terminate the thrust. With liquid propellants this is simple; control the rate at which the propellants are pumped into the combustion chamber. With solid propellants, the combustion rate is controlled by the geometry of the propellant, which cannot be changed on command. However, the thrust can be terminated at any time by rapid depressurization; that is, explosively opening the combustion chamber. Fortunately, there are many applications in which the combustion rate can be preplanned, so abrupt termination is not necessary.

4

MILESTONES IN JET PROPULSION

The turbojet engine was invented in 1930, as described in Frank Whittle's British patent. It could propel an airplane at high speed without a propeller. Development was slow, and the first flight of an airplane propelled by Whittle's engine was not until May 1941. See Figure 11. Meanwhile, Dr. Hans von Ohain in Germany started work on a jet engine in 1935, which led to a flight of a Heinkel HE 178 in August 1939 piloted by Erich Warsitz. Subsequent development was slow, because both England and Germany were fully occupied by World War II.

FIGURE 11. WHITTLE'S ENGINE

Both Whittle's and von Ohain's engines used centrifugal flow compressors, which presented a large frontal area. It was soon realized that axial flow compressors would be better, since they provided a smaller frontal area, and also could provide more stages of compression. Anselm Franz of Austria was the first to design a jet engine using such a compressor, and it was flown in 1944, propelling a Messerschmitt Me 262 fighter plane. This was the first jet fighter, and it received limited use in the final phase of World War II. It was faster than any other plane at the time, but the war ended before it could be used on a large scale. About 1400 of these planes were produced. See Figure 12.

FIGURE 12.
MESSERSCHMITT ME 262 FIGHTER PLANE

After the war, the military in many countries began developing jet planes, both fighters and bombers. By the end of the 1940's, most planes being produced were turbojet planes. In 1947, Boeing introduced the B-47 Stratojet, which was propelled by six engines, each producing 3750 pounds of thrust. In 1954, this was rendered obsolete by the B-52 Stratofortress, which had a much longer range, 8800 miles, and much greater load capacity. It can fly at an altitude of 50,000 feet. It is propelled by eight Pratt and Whitney engines each producing 17,000 pounds of thrust. A total of 744 units were produced, and, amazingly, the plane is still in service after more than 50 years.

FIGURE 13. THE B52 STRATOFORTRESS

The first commercial use of a jet plane was in 1952, when the British de Havilland Comet was put into service. It offered service from London to Johannesburg at 480 miles per hour, and was very popular. Unfortunately, in 1954, two Comets broke up in midair. Comet flights were then discontinued, and a redesign was undertaken. The new Comet was not ready to fly until 1958, and by that time a number of jets suitable for commercial use had appeared: the Tupolev 104, the Boeing 707, and the Douglas DC-8.

FIGURE 14. THE DE HAVILLAND COMET

Afterward, numerous models of commercial jets appeared, largely manufactured by Boeing and Airbus. Especially notable was the Boeing 747, the first wide-body jet, which was larger and longer-range than any predecessors. It weighed 450 tons and could fly at Mach 0.85. Its power came from four engines, either Pratt and Whitney or General Electric, each producing 63,200 pounds of thrust. It could carry up to 500 passengers. Over 1400 of these planes have been produced to date. See Figure 15.

FIGURE 15. THE BOEING 747 JET PLANE

The Airbus A-380, first flown in April 2005, is even larger, weighing 615 tons at take-off, and can carry 555 passengers in three classes, or 853 passengers in economy class. It has a range of 8000 miles. A freight version is able to carry 150 tons. Its cabin has 50 % more floor space than the Boeing 747. As of January 2010, 41 units have been built.

FIGURE 15A. THE AIRBUS A-380 JET PLANE

All jets mentioned above were subsonic; they flew at speeds up to Mach 0.88. At one time there was concern about a "sound barrier" at Mach 1. Fighter planes when diving could exceed Mach 1, but sustained supersonic flight was not demonstrated until 1947, when Chuck Yeager flew the Bell X-1, a rocket-propelled experimental plane, through the sound barrier.

FIGURE 16. THE BELL X-1 ROCKET PLANE

The first supersonic fighter plane was the F-100 Super Sabre, built by North American. It was powered by a Pratt and Whitney J57 engine producing 10,000 pounds of thrust, which could be increased to 16,000 pounds when an afterburner was used. It first flew in 1953, and was used until 1971 by the U.S. Air Force. Its top speed was 822 miles per hour, well over the speed of sound. A total of 2300 units were delivered. See Figure 17.

FIGURE 17. THE F-100 SUPER SABRE FIGHTER PLANE

The first United States—built supersonic bomber, the B-58 Hustler, could fly at Mach 2 at 64,000 feet. It first flew in 1956, and since then 116 units were produced. It had a delta wing, and was propelled by four engines producing 9700 pounds of thrust each. It had a small bomb capacity, having been designed to carry a single nuclear bomb.

FIGURE 18. THE B-58 SUPERSONIC BOMBER

The next important U.S. bomber was the XB-70 Valkyrie, which could fly at Mach 3.1. It first flew in 1964, but only two were built before the program was cancelled. It was followed by the B1-A, which flew at Mach 2+ in 1974. It was succeeded in 1986 by the B1-B, called the Lancer, which flew at Mach 1.25 at 60,000 feet. It was also capable of flying fast at sea level (below the radar). It was propelled by four General Electric engines each producing 14,600 pounds of thrust. When afterburners were turned on, the thrust would double.

FIGURE 19. THE B1-B LANCER BOMBER

Dozens of additional supersonic fighters and bombers have been built, but will not be listed here. However commercial supersonic planes were slower to appear. In 1968 the Rusian supersonic plane Tupolev TU 144 was put into service. It could fly at Mach 2.35, and could carry up to 140 passengers. However its reliability and fuel efficiency were not satisfactory, and it was withdrawn from service in 1978 after about a hundred commercial flights. In 1969 a prototype Concorde, jointly made by Britain and France, was test-flown, and in 1976 it went into scheduled commercial service, flying at Mach 2. It could carry one hundred passengers. In subsequent years, eighteen Concordes had flown. See Figure 20.

FIGURE 20. THE CONCORDE

The problem of sonic boom caused by supersonic overflights was troubling, and the Concorde was restricted to supersonic flying over water, with subsonic flying over land. (Sonic boom is a transient pressure wave reaching earth when a supersonic plane flies overhead.)

The Concorde was not a commercial success, largely because it could not carry enough passengers. Also, its fuel costs and maintenance costs were high. In July 2000 a Concorde had a disastrous accident when taking off from a Paris airport. The Concorde was withdrawn from service in 2003. At present, there is no scheduled supersonic plane flying.

Let us turn now from turbojet to ramjet propulsion. This offers possibilities of much higher flight speeds. At Mach 10, one could fly halfway round the earth in less than two hours. This is in the future. Another potential use for ramjets is as a first stage for a space vehicle.

Today, the only practical use of a ramjet is as an anti-aircraft missile.

In 2004, NASA broke ground by launching an unmanned research vehicle, designated X-43A, propelled by a supersonic combustion ramjet (scramjet) fueled with hydrogen. It had a successful flight, reaching Mach 9.6. See Figure 21.

FIGURE 21. NASA's X-43A RESEARCH VEHICLE, WITH SCRAMJET

5

HOW ARE ROCKETS USED?

The most obvious applications are for space exploration and in military weapon systems. Before discussing these, however, there are a number of other uses for rockets.

One example is meteorological. A relatively small solid-propellant rocket a few inches in diameter and a few feet long is widely used to monitor upper-atmosphere temperature and wind. The purpose is to predict weather changes.

The Coast Guard uses rockets to throw lifelines to ships.

Rockets are used to assist take-off of airplanes.

Rockets launch satellites which perform many functions: telephone, radio, and television communication; observation of the earth for meteorology, agriculture, geology, the ozone layer, etc.; sending signals for global positioning (GPS); and espionage (spy in the sky).

The use of satellites in earth science deserves further discussion. These studies clarify risks from global warming, earthquakes, and hurricanes, and permit tracking the condition of forests, fisheries, and water resources. Satellite-borne instruments like high-resolution photography, radar and lasers, have revolutionized earth and climate science. For example, changes in sea level can be tracked. Movements of the earth due to earthquakes can be followed. Rainfall and moisture in the air can be measured, as well as the average temperature of layers of the atmosphere. The knowledge of frequent changes in location of the jet stream is valuable to airlines.

The International Space Station (reachable only by rocket) provides a unique environment for studying phenomena in the absence of gravity.

The location of powerful telescopes (e.g., the Hubble) in space has proven invaluable to astronomers. By being able to detect very distant galaxies, not visible from earth, and measuring their velocities, they can make deductions about the age of the universe and its rate of expansion since the big bang fourteen billion years ago. Other orbiting instruments can measure a variety of radiations from space other than visible light.

Rockets are used for signaling when rescue is needed.

Ejection from an aircraft in trouble may use rockets.

How else could we celebrate the Fourth of July?

The use of rockets for space exploration is widely known. Indeed, there is no alternate way to leave the earth. But rockets are not only used for primary propulsion; they are used for attitude control, for stage separation, to impart spin, as igniters of large solid-propellant rockets, for braking, etc.

It may be appropriate to ask if space exploration serves any useful purpose. It seems to me that a successful search for present or past life on other planets would have profound effect on biology. At present we know that every current or past animal or plant on earth has a degree of DNA commonality. We have no way to predict what the DNA situation would be for any non-earth life form. The question of how life originated on earth would seem to be of crucial importance.

Also, it has been suggested that a large meteorite may some day strike the earth with devastating results, as has apparently happened several times in the past, causing so-called extinctions. The most recent was 65 million years ago, which is believed to have wiped out the dinosaurs. Any possibility of diverting such a meteorite would require space-travel capability.

As for military applications, it is well known that rockets are used to destroy airplanes in flight. There are both ground-to-air and air-to-air rockets. A research program has long been under way to see if rockets can be used to intercept incoming long-range ballistic missiles, so far with only very limited success. Rockets can also be used to attack land-based targets, such as tanks, ships, or forts. As was previously mentioned, a rocket is an ideal way of carrying a nuclear bomb.

A military rocket might differ significantly from a rocket used for civilian purposes, for three reasons. First, a military rocket ideally should be capable of being stored for years, and then launched instantly. Accordingly, solid-propellant rockets are preferred by the military rather than those using liquid propellants, especially cryogenic liquids. Secondly, consider rockets to be launched from a submarine. In this case, leakage of a toxic or explosive vapor would be unacceptable, which argues for a solid propellant. Thirdly, for many military applications it is desirable that the rocket exhaust be smokeless, so as not to reveal the launch location. This would limit the choice of propellants for such applications.

6

DEVELOPMENT OF ROCKETS FOR SPACE TRAVEL

Ever since Sir Isaac Newton formulation his laws of motion and of gravitation, it has been possible to calculate the velocity that must be achieved for a rocket, or any other object projected upward, to escape the earth's gravitational field rather than falling back to earth. The calculation is presented in Appendix B, for those with some knowledge of physics and mathematics. In summary, the calculations show that a velocity of five miles per second is needed to hurl an object into earth orbit, and a velocity of seven miles a second is needed to escape the earth and go to the moon or Mars.

A few words may be said about the concept of an earth orbit. If a heavy weight attached to a string is whirled about, while the string is held, the weight will continue to move in a circle until air resistance slows it down. The tendency of the weight to fly away is balanced by the tension in the string. For a spaceship in orbit, moving at five miles per second, there is no string, but there is a gravitational force from the spaceship to the center of the earth. Thus, an orbit. By the same principle, the moon orbits about the earth, and the earth orbits about the sun. An object orbiting the earth, if completely above the earth's atmosphere, will continue to orbit for many years. The time to complete each orbit depends on how high above the earth it is. An orbit just above the earth's atmosphere (say, 100 miles up) will occur in one and a half hours. An orbit 22,000 miles up will occur in 24 hours; this is called a geosynchronous orbit.

It is possible to calculate (again shown in Appendix B) that even the most ideal rocket, using the best known propellants, could never accelerate to the needed speed of five miles per second. The rockets that existed at the beginning of the twentieth century could never reach even two miles per second. Accordingly, scientists at that time concluded that rockets could never be used for a journey into space.

In 1865 the French novelist, Jules Verne, wrote a novel, "From the Earth to the Moon", which describes a manned trip to the moon (from Florida, no less). Launching in an easterly direction from near the equator takes advantage of the earth's rotational speed about its axis, which is close to three-tenths of a mile per second. Of course, another 6.7 miles per second is needed to reach the escape velocity of seven miles per second, and probably Jules Verne was aware that this was far beyond the capability of any rocket. Therefore he assumed the use of a gigantic cannon nine hundred feet long. This could never have worked, for two reasons: a cannon could never produce the necessary muzzle velocity; and even if it could, no human could survive the extreme acceleration.

However, the possibility of using rockets for space travel was clarified in 1903, when a Russian teacher and inventor, Konstantin Tsiolkovsky, published his calculations showing how the concept of staged rockets would permit very high velocities to be reached. For example, a very large rocket could be used as a first stage, which might be able to accelerate a payload to 2.5 miles per second. The payload would consist of a smaller second rocket, which could then accelerate from 2.5 to 5 miles per second. The payload of the second rocket could be an even smaller third rocket, and when this rocket, the third stage, was fired, it could accelerate its payload from 5 to 7.5 miles per second. Accordingly, if one started with a sufficiently large rocket, and had a sufficient number of stages, any desired final velocity could be reached. Tsiolkovsky also pointed out that higher velocities per stage could be achieved by using liquid propellants instead of solid propellants. However, in 1903 no one had ever built a powerful liquid-propellant rocket.

Tsiolkovsky's publications in Russia did not attract much worldwide attention. However, in 1922 a German physicist, Hermann Oberth, wrote his doctoral thesis, which explained how rockets could be used for space travel, in which he duplicated Tsiolkovsky's treatments, probably unknowingly. His thesis was rejected as "utopian". However, his study of rockets continued for many years. He ultimately obtained his doctorate. In 1928 he served as scientific consultant on a pioneering German film, "Frau im Mond" (The Woman in the Moon) directed by Fritz Lang. This film was important in popularizing rocket science.

Meanwhile, in the United States, Professor Robert Goddard of Clark University (Worcester, Massachusetts) was aware that staging and liquid propellants were keys to space travel, and in 1922 he was apparently the first to design a liquid-propellant rocket. The first flight test of one of his rockets, in 1926, was not very successful, but he persevered, with very little support. His best result was a rocket that reached a speed of 741 miles per hour. and reached an altitude of 1.7 miles. See Figure 22.

FIGURE 22. GODDARD AND ONE OF HIS EARLY ROCKETS

In 1929 Professor Oberth, at the Technical University of Berlin, launched his first liquid-propellant rocket. He was helped by a group of his students, including Werner von Braun. Throughout the 1930's, the German group continued to develop rockets, having space travel in mind rather than military applications. Simultaneously, a group in Russia headed by Sergei Korolev experimented with rockets. Korolev's work was supported by the Russian government until 1938, until a dispute arose; Korolev wanted to continue working with liquid propellants, good for space travel, while the government wanted him to shift to solid propellants, since the latter were thought to be more practical for military purposes. The dispute was settled by shipping Korolev and co-workers to a Siberian prison camp.

Then, in 1939, World War II began, and Werner von Braun, now with a doctorate, became head of a large group at Peenemuende, Germany which developed the V2 rocket. This work was supported by the German government, and was the first large-scale effort to build a practical rocket.

The V2 rocket program was successful, the first launch occurring in October 1942. A brief description of the rocket is presented. It weighed about 14 tons at launch, and carried a payload of about a ton of explosive. Its fuel was a 75-25 alcohol-water mixture, with an oxidizer of liquid oxygen. It was gyroscopically stabilized until burn-out (sixty five to seventy seconds after launch), after which it moved like an artillery shell, with no guidance. Accordingly it was not very accurate, and at its maximum range of 225 miles it could deviate from its intended target by four to eleven miles. However, when the target was sufficiently large, like the city of London, it could be hit. In 1944 and early 1945 about 500 of these rockets, fired from Germany or the channel coast, hit London. About 3000 were fired in all, some missing the target, some at other targets and some misfires. The top speed achieved by this rocket was about one mile per second.

FIGURE 23. THE V2 ROCKET, SHOWING COMPONENTS

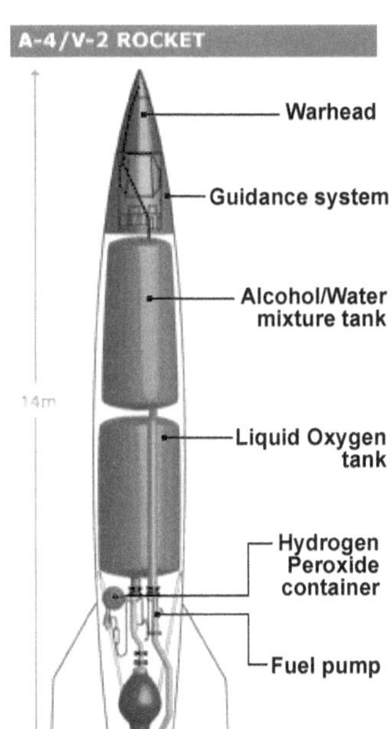

The war ended in 1945, the same year that the nuclear bomb was first set off. It became apparent to the military in all nations that the use of a rocket somewhat improved over the V2, carrying a nuclear warhead, could threaten a target anywhere in the world. It would appear not to be interceptable. Accordingly, the United States brought von Braun and some of his fellow workers to the United States and supported their rocket development, while simultaneously supporting other programs in industry and government laboratories.

Meanwhile, in the USSR, Sergei Korolev was released from imprisonment and put in charge of the Soviet long-range rocket program.

In the 1950s and 1960s, the U.S. and the U.S.S.R. engaged in an arms race, with virtually no restraint on the spending of funds to develop long-range rockets. By 1957, both countries were able to put earth satellites into orbit (requiring acceleration to about five miles per second), and trips to the moon began in the 1960s. How was the V2 rocket upgraded so much? In the first place, it was known that there were more powerful liquid propellants than what was used in the V2. In the second place, the concept of staging could be applied. In the third place, it was possible to develop much more sophisticated guidance and control systems than the V2 had. Furthermore, the military had a preference for solid propellants, and these were significantly improved, although not to the point of being quite as powerful as liquid propellants. Combinations of liquid and solid propellants were considered, and used in the Space Shuttle program.

In the 1950s, it was realized that there was a major problem of a high-speed spacecraft re-entering the earth's atmosphere after a mission in space. This re-entry problem is the result of aerodynamic heating. Unless something were done to prevent it, the temperature of the front surface of the missile would become so high during the several minutes of re-entry that the heat would melt and destroy the missile. The kinetic energy of a missile moving at five miles per second or more has to be dissipated somehow as the missile slows. Of course, the missile could be substantially slowed down before re-entry by firing a braking rocket (a retro rocket), but the weight of such a rocket and its propellants adds substantially to the problem of launching the spacecraft in the first place. If nothing is done to prevent it, the energy will be dissipated by heating the skin of the spacecraft. Some part of the heat will be radiated away, but the majority will be conducted into the spacecraft and destroy it. Note that stony or metallic meteorites frequently enter the earth's atmosphere at high speed and are vaporized by the heat (shooting stars).

Much engineering effort was directed to this problem in the 1950s and 60s, and several solutions were found. One successful approach was to use ablative cooling. This involves coating the vulnerable surface with a layer of plastic, perhaps reinforced with fibers, which, upon being heated, would partly gasify and partly turn into a char. The gas generated would tend to push away exterior very hot air, and the remaining char would radiate heat away while insulating the spacecraft. This ablative cooling approach worked very well, but, in order to re-use a spacecraft, a new layer of ablative material would have to be applied each time. Hence, an alternate approach was developed using tiles made of silica foam that were attached to the exterior of the spacecraft. This worked very well, except some of the tiles would occasionally fall off. This happened once to the Space Shuttle, causing a fatal accident.

As of now, the Space Shuttle is the most sophisticated, and proven, spacecraft we have. The first Shuttle launch was in 1981. It does not seem fruitful to describe each of the various spacecraft used between 1957 and 1981, except to say that they became increasingly sophisticated, culminating in the Shuttle. A brief description of the Shuttle, shown in Figure 24, is given.

FIGURE 24. NASA'S SPACE SHUTTLE

It is large! The lift-off weight is 2200 tons. It stands 184 feet high on the launch platform. Its engines produce about six million pounds of thrust during the first two minutes of flight. The engines consist of a group of three liquid-propellant engines fueled by liquid oxygen and liquid hydrogen, and two large, detachable solid-propellant rockets flanking the main engines. During the first two minutes, the solid propellants provide about 80 percent of the thrust and the main (hydrogen-oxygen) engines provide the other 20 percent. After two minutes, the solid-propellant engines are detached, fall to the ocean, and are recovered. The main engines continue to operate for several minutes more, until the spacecraft achieves 4.81 miles per second, at which time the engines are separated from the Orbiter, which contains the crew and payload. The Orbiter goes into orbit around the earth at a selected altitude, roughly 120 miles up.

The Orbiter generally has a crew of seven astronauts, and carries appropriate equipment and supplies for each mission, which may be to help construct the International Space Station, or to install or service the Hubble telescope, or to put communication satellites into orbit, or to perform scientific experiments. To date, the Space Shuttle has made over 20 trips to the International Space Station. It is now considered almost obsolescent. It will probably be replaced by a new spacecraft in a few years, the design of which has not been finalized.

Rockets capable of putting payloads into orbit have been built not only by the United States and the U.S.S.R., but also by the European Space Agency, China, France, India, Israel, and Japan. There is a worldwide interest in rocketry. To date, about 400 people have orbited the earth at least once.

To summarize, over fifty years elapsed between the emergence of the concept of rocket staging and the first orbiting vehicle. Progress was very slow for the first thirty of these years, because no vigorous government or industrial support was present. Then, World War II led to the V2 rocket and the nuclear bomb, and much effort was directed subsequently toward rocketry. Twelve years after the war ended, both the USSR and the US demonstrated capability of putting payloads into earth orbit. The present status is that we can send space vehicles to any of the planets, but only with the use of large and expensive launching procedures. The possibilities for future development of significantly better propulsion techniques will be discussed in a subsequent chapter.

7

SELECTION OF JET FUELS AND ROCKET PROPELLANTS

We are all familiar with gasoline as used for the reciprocating engines of automobiles. Two characteristics must be considered: the octane number of the fuel (to avoid knocking) and the volatility (so that the fuel can vaporize completely in the carburetor). For turbojet engines, the octane number of the fuel is irrelevant, and the vaporization characteristics also do not matter, because the fuel in liquid form is sprayed into the combustion chamber and easily vaporizes because of the high temperature in the combustion chamber. Accordingly, virtually any hydrocarbon mixture may be used as a turbojet fuel, which reduces its cost relative to high-octane gasoline.

It is desirable to select a hydrocarbon mixture with as high a density as possible, so as to reduce the size of the fuel tank, and hydrocarbons with densities ten or fifteen percent higher than gasoline for automobiles are used. It is also desirable to have a jet fuel with less volatility than gasoline, for safety reasons in case of spills. Today, the U.S. Air Force uses JP-8 while commercial jet aircraft use JET-A. These are similar, and are essentially kerosene. They cannot be ignited with a match at normal ambient temperatures.

There is very little practical experience with ramjets, which may operate at speeds of Mach 2 to Mach 10 or higher. At the lower end of this range, the fuel is not critical. Any type of hydrocarbon fuel could be used. At the higher Mach numbers, two problems arise regarding the fuel selection. At Mach numbers of 5 or above, aerodynamic heating of the vehicle becomes serious. If the fuel is to be circulated to cool critical areas, it is desirable that the fuel be thermally stable and not break down to form carbonaceous material while being circulated. This consideration eliminates most hydrocarbons. Methane is the most stable hydrocarbon, and is a candidate. The other candidate is hydrogen, which is stable when heated even above 2000 degrees Fahrenheit. Methane has an advantage over hydrogen because of its much higher density in liquid form, about 6.6 times as great.

However, the second problem at high Mach numbers, with supersonic velocity through the combustion chamber, is the need for very rapid combustion in the limited space and limited time available. The rate of combustion is controlled by the chemical kinetics of the oxidation. Hydrogen has a considerable advantage over methane in this regard, having a lower spontaneous ignition temperature and greater reactivity.

Rockets function by burning something in a combustion chamber at high pressure, and then expanding the hot combustion products through a nozzle. In the nozzle, a large fraction of the thermal energy in the combustion products is converted to kinetic energy. The high-velocity gases issuing from the nozzle create thrust.

The magnitude of the thrust depends on the chemical nature of the propellants, on the pressure in the combustion chamber, and to a lesser extent on the external pressure. When comparing various propellant combinations, it is convenient to assume that the combustion chamber pressure has a specific value, 1000 psi for instance, and the external pressure is the normal atmospheric pressure. Then, the only difference between the propellant combinations is the chemistry.

Each rocket propellant system may be characterized by a single number, the specific impulse, as long as the comparison is made at the same combustion chamber pressure and external pressure. From knowledge of the specific impulse and the ratio of rocket mass before and after the burn period, a formula is available to calculate the increase in velocity of the rocket. See Appendix B for mathematical details. Figure 25 is a diagram showing the relation between specific impulse, mass ratio, and velocity increase.

FIGURE 25. DEPENDENCE OF ROCKET VELOCITY INCREASE ON SPECIFIC IMPULSE AND MASS RATIO

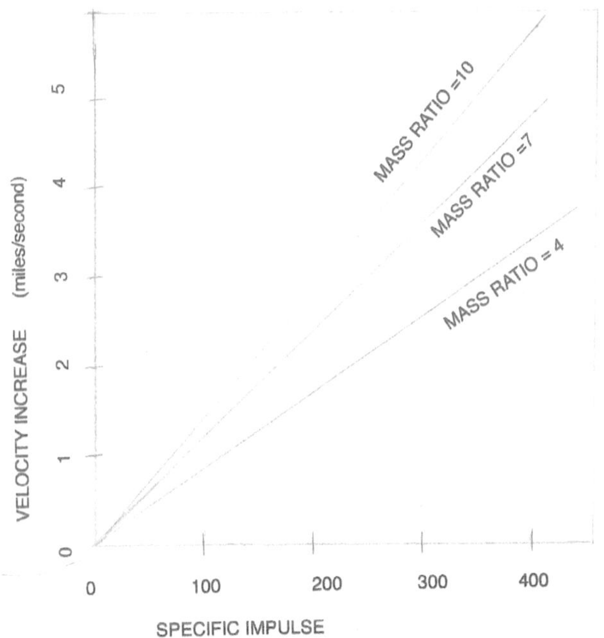

There is another formula that permits calculation of the exhaust velocity of the rocket in terms of the specific impulse. Multiply the specific impulse, in seconds, by 0.0061 to obtain the exhaust velocity, in miles per second. For example, the specific impulse of a hydrogen-oxygen rocket is about 389 seconds, and the corresponding exhaust velocity is 2.37 miles per second. From the formula provided in Appendix B, a hydrogen-oxygen rocket with the propellant mass 75 percent of the total initial mass would reach a velocity of 3.3 miles per second. Accordingly, a two-stage hydrogen-oxygen rocket with a mass ratio of 4 (100/ 25) can achieve a velocity of twice 3.3, or 6.6, miles per second. Since orbital velocity is about 5 miles per second, such a rocket is capable of reaching orbit. Another stage would permit escape from the earth.

While hydrogen-oxygen is the most powerful practical rocket propellant combination known, there is difficulty in handling and storing these liquids, which boil at temperatures far below ambient. So-called cryogenic techniques are used. There are other rocket applications than space travel that could use other, more storable propellants. The choice of one of these alternates will largely depend on how much specific impulse is needed. Let us discuss specific impulse in detail.

Specific impulse is defined as the thrust force produced by a rocket during firing, divided by the mass rate of combustible consumption. It may be calculated from the known chemical constituents of the combustibles, or it may be measured directly during a firing of a rocket on a test stand. In general, the measured value will be lower than the calculated value, for several reasons. The combustion may not be complete. The recombination of dissociated species in the divergent section of the nozzle may not be complete. The flow out of the rocket nozzle may be somewhat divergent rather than strictly parallel to the major axis of the rocket. The friction at the nozzle walls may slow down the flow. Some of the heat generated by combustion may pass into the walls of the combustion chamber or the nozzle rather than remaining in the gases. If the gases contain solid or liquid particles, such as aluminum oxide, the particles may not accelerate in the nozzle as fast as the gases. In spite of all these reasons, the measured specific impulse is usually only five or ten percent lower than the calculated value.

Since the calculations come so close to predicting the actual performance, they provide an important guide in how to select the proportions of the propellant ingredients. A propellant combination usually consists of a fuel and an oxidizer. For liquid propellants the oxidizer is generally liquid oxygen, and, for best performance, the fuel is liquid hydrogen. A possible alternate to liquid hydrogen is a hydrocarbon liquid (RP-1). Since liquid hydrogen has an extremely low density, much smaller space is required to store a hydrocarbon liquid, which has a density eight times that of liquid hydrogen. For some missions, this space saving is of decisive importance.

For many rocket applications, a solid propellant is preferred over a liquid propellant combination, even if it doesn't have as high a specific impulse. For solid propellants, the oxidizer may be a perchlorate or nitrate crystal, or a nitrated organic material. The fuel may be any of a large choice of plastics, chosen to act as a binder and provide mechanical strength to the propellant "grain", which is a single piece of selected geometry. The "grain" must be strong enough to withstand the force acting during acceleration of the rocket. Very often, the solid propellant contains ten to twenty percent of aluminum powder, which produces a higher flame temperature.

The calculations show that the higher the temperature of the combustion products and the lower their mean molecular weight, the greater the specific impulse. (Hydrogen has the lowest molecular weight of any element, so a high proportion of hydrogen in the exhaust is desirable.) In the case of a hydrogen-oxygen rocket, the relative proportions of hydrogen and oxygen fed into the combustion chamber must be decided. Any chemist knows that two molecules of hydrogen combine with one molecule of oxygen to form two molecules of water, after combustion. It turns out that combining hydrogen and oxygen by this prescription results in a higher flame temperature than if the proportions represented an excess of either hydrogen or oxygen. The higher the flame temperature, the higher the specific impulse, other factors being the same. But a low average molecular weight of the combustion products also favors high specific impulse. It turns out that higher specific impulse is obtained by using four molecules of hydrogen with each molecule of oxygen. The exhaust gas will then consist of two molecules of hydrogen for each molecule of water vapor.

Let us consider a typical solid propellant, a mixture of small ammonium perchlorate crystals, a plastic binder, and aluminum powder. The optimum composition of such a mixture may be approximated by assuming that just enough oxidant is provided to oxidize all the aluminum to aluminum oxide and to oxidize all the carbon in the organic binder to carbon monoxide (not carbon dioxide). Most of the hydrogen from the ammonium perchlorate and the plastic binder will not be combined with oxygen, but will be in the exhaust gases as free hydrogen and hydrogen chloride. (The calculated specific impulse of such a mixture is about 250 seconds, compared with 389 seconds for the hydrogen-oxygen rocket.)

The specific impulse of any combination of rocket propellant chemicals can be accurately calculated, by a procedure discussed in Appendix B. As previously mentioned, this theoretical specific impulse will always be slightly higher than the actual specific impulse. Nevertheless such calculations are extremely useful in selecting the chemical nature and proportions of the propellant combination.

In the development of any new rocket, the actual specific impulse is always measured on a thrust stand in a test firing, and the result compared with the calculated value. How does one measure specific impulse? As far as is known, the first accurate measurements were done in 1915 by Professor Robert Goddard, using a ballistic pendulum. This ballistic pendulum technique was first used by Benjamin Thompson (Count Rumford) more than a century previously to determine the speed of a projectile fired from a cannon. From the degree of rise of a heavy pendulum when a projectile was fired into it, he could calculate the speed of the projectile. Goddard modified this by attaching a rocket to a heavy pendulum and then igniting it and observing how high the pendulum rose. A better procedure was soon devised in which the force exerted on the test stand was measured continuously by observing the compression of a spring or, more accurately, by a strain gauge attached to the test stand. See Figure 26.

FIGURE 26. A ROCKET BEING FIRED ON A TEST STAND

When a firing is done on a test stand, the exhaust gases encounter atmospheric pressure. When a rocket is fired high above the earth, the ambient pressure is much lower, and is essentially zero in deep space. This lower exhaust pressure results in greater specific impulse, which can be calculated.

Let us consider the burning process in the combustion chamber. In the case of liquid propellants, the fuel and oxidizer are pumped into the combustion chamber and sprayed through nozzles. The combustion causes the pressure in the combustion chamber to rise to a high value, perhaps 1000 psi, and the pumps must produce pressure higher than this. The temperature is so high that the combustion occurs very rapidly. The thrust can be varied or terminated by controlling the pumps. A problem that sometimes is encountered in the course of engine development is combustion instability. There are several possible modes of instability, and engineering solutions have been found for all of them. To preserve the engine, and especially the nozzle, from the high temperature of the gases, it is customary to provide cooling by circulating one or both of the cryogenic liquids through appropriately located coils.

For solid propellants, it is a different story. The combustion occurs in a thin zone just at the surface of the "grain". (The "grain" is the term used to denote the continuous piece of propellant in the combustion chamber.) The burning rate, which may be expressed in millimeters per second, is the rate at which the combustion zone regresses into the "grain". It generally increases with increasing combustion chamber pressure. The rate of generation of combustion products is proportional to the amount of exposed surface of the "grain".

The designer of the rocket must select a grain geometry that provides enough surface to give the desired combustion rate. Also the thickness of the "grain" must be such that burnout occurs at the desired time. It is usually desirable for the rate of gas generation to be more or less constant during the burning period. One way to accomplish this is with an end-burning geometry. If this does not give a high enough gas generation rate, then we may select a geometry with the burning occurring on the inside of a cylindrical hole in the "grain". However this design is not ideal because the combustion rate and the chamber pressure both increase during the burn, as the diameter of the hole becomes larger. Another design overcomes this problem; combustion occurs on the star-shaped projections in the hole through the grain, as shown in Figure 27. As the burning continues, the projections are burned away, and at the end of the burn they have disappeared. It can turn out that the rate of gas generation is constant during this burn.

FIGURE 27. END VIEW OF GRAIN GEOMETRY FOR CONSTANT BURN RATE

Many other variations are possible. For example, it may be desirable to have a high combustion rate for the first half of the burn period (booster phase), and a slower combustion rate for the second half of the burn period (sustainer phase). This can be accomplished with an appropriate grain design.

There is another tool available to the designer. The chemical content of the propellant can be varied by introducing a small amount of additive to the propellant which acts as a catalyst or inhibitor, increasing or decreasing the burning rate, perhaps by a factor of two. The burning rate also depends on the fineness of grinding of the oxidizer crystals. Much greater increase in burning rate can be obtained by mixing short strands of copper or silver wire into the propellant. For end-burners, the wires may be continuous in the longitudinal direction.

Various techniques are used to ignite a solid-propellant rocket. A large rocket may be ignited by electrically firing a small rocket so located that its exhaust passes over the exposed grain surface. This small rocket may contain boron, potassium nitrate, and a binder. Alternately it may contain magnesium and Teflon.

Solid propellant compositions often include ten or twenty percent of fine aluminum powder. When this aluminum burns, producing aluminum oxide droplets, a much higher flame temperature is produced, which leads to a higher specific impulse. It so happens that there is another beneficial effect of aluminum; if there is a tendency for an acoustic oscillation to occur in the combustion chamber, with possibly destructive effects, the presence of the aluminum oxide droplets in the combustion gases acts as a damper to suppress the oscillations.

Some examples of solid propellant ingredients may be mentioned. The most common oxidizer is ammonium perchlorate, in the form of crystals. There is an advantage in using a bimodal distribution of crystal sizes, so that smaller crystals may fit between the larger crystals. Then, when the oxidizer is mixed with the binder and aluminum powder (before polymerization), there is greater fluidity of the mix, and a higher oxidizer-binder ratio may be worked with. An alternate to ammonium perchlorate is ammonium nitrate, which may be used when a lower combustion temperature is desirable, as in a gas generator. All the propellants using crystalline oxidizers are referred to as composite propellants.

As a result of the aluminum addition, the exhaust will be a very smoky and hence easily visible plume. This may be unacceptable in some military uses. In such a case, an entirely different type of solid propellant may be used, called double-base. The "grain" consists of nitrated cellulose impregnated with nitroglycerine. Double-base propellants have the disadvantage that they slowly decompose after being stored for several years.

Any of a large number of plastics may be used as binders for composite propellants. Examples are polybutadiene-acrylic acid-acrylonitrile and polyurethane.

The "grain" is generally polymerized after being cast to form the desired shape. It may either be case-bonded or cartridge-loaded.

Approximate values of specific impulse for the various types of propellant systems mentioned to date are:

Liquid hydrogen/liquid oxygen	389 seconds
Hydrocarbon/liquid oxygen	300 seconds

Composite solid propellant, aluminized	250 seconds
Double base	225 seconds

The discussion to date has been limited to propellants for primary propulsion. As discussed in the next section, there is also a need for smaller rockets for guidance and control of the primary spacecraft, or to permit it to perform maneuvers when in orbit, for example. For such rockets, cryogenic fuels might be used during the primary thrust period, by borrowing fuel and oxygen from the main tanks, but, after the primary thrust period, there is an obvious advantage in using readily storable fuels. It is not critical if such fuel combinations do not have the maximum possible specific impulse.

Liquid propellants are preferred over solid propellants for such smaller rockets, because of their on-off-on capability. (It is possible to terminate the firing of a solid-propellant rocket by sudden rapid pressure drop in the combustion chamber, which can be accomplished by explosive opening. However restart would not be an option.) The types of liquid propellants in use may be divided into monopropellants and bipropellants.

A monopropellant is a liquid that can decompose exothermically when entering the combustion chamber, without requiring an oxidant. This is possible because the substance can decompose while liberating heat. While many chemicals are capable of this behavior, the most frequently used are hydrazine and hydrogen peroxide. Hydrazine decomposes exothermically to form the gases ammonia, nitrogen, and hydrogen. Hydrogen peroxide decomposes exothermically to form water vapor and oxygen.

A great variety of bipropellant storable liquid combinations are possible. The United States mainly uses monomethylhydrazine and nitrogen tetroxide, for example on the Space Shuttle, while Russia and China mainly use unsymmetrical dimethylhydrazine and nitrogen tetroxide. Both these combinations have the advantage that they are hypergolic; that is, they spontaneously ignite when the fuel and the oxidizer come in contact. Hence no igniter is needed.

One other propellant combination that may be mentioned is used in the hybrid rocket, which involves both a solid and a liquid as propellants. This usually consists of liquid oxygen as the oxidant and an inert solid fuel grain, often a polymer such as polybutadiene. Instead of liquid oxygen, it is also possible to use other liquid oxidants and other solid fuels. The advantages of a hybrid system are: safety during storage or operation with no possibility of explosion; start-stop-restart capability; relatively low system cost; higher specific impulse than solid-propellant motors; high density-specific impulse relative to hydrogen/oxygen; and the ability to smoothly change thrust over a wide range. The disadvantages are: specific impulse may undesirably vary during operation; the density-specific impulse is less than that of solid-propellant motors; some fuel sliver may be retained at the end of burn. See Figure 28.

FIGURE 28. A HYBRID ROCKET ENGINE

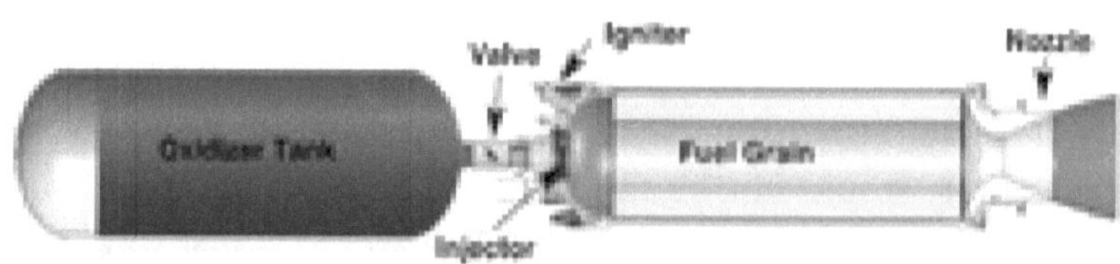

No large-scale hybrid rocket has been proven yet. While the specific impulse of the liquid oxygen-polymer combination has been calculated to be over 300 seconds (better than solid propellants), it is possible to consider more exotic oxidizers than oxygen, such as fluorine or fluorine oxide, and more exotic fuels, such as lithium hydride or aluminum hydride. Such combinations may yield specific impulses as high as 380, but no such exotic system has ever been tested in flight.

A search has gone on for many years, seeking a superperformance propellant combination substantially superior to hydrogen and oxygen. This can be done with a computer, calculating the specific impulse of each candidate system, as described in Appendix B. The most powerful combination that has been found is a combination of liquid hydrogen, liquid fluorine, and powdered beryllium, which gives a calculated specific impulse 23 percent higher than hydrogen-oxygen. However, it is unlikely ever to be used, because both fluorine and beryllium oxide are highly toxic.

8

GUIDANCE AND CONTROL OF ROCKETS

Why would a rocket require guidance and control? There are at least six reasons. First, it would have to be guided if we want it to follow a specified trajectory. Second, control may be needed to maintain stability during flight. Third, a guidance system may have to compensate for forces throwing the rocket off course, for example crosswinds or forces occurring during stage separation.Fifth, in order for a rocket to dock with a satellite, or to make a soft landing on the moon, there must be a means to cause small, gentle motions. Sixth, military rockets are often intended to strike a target, perhaps a moving target trying to evade the rocket.

Clearly, then, almost any rocket (except a meteorological sounding rocket which is shot straight up, makes its measurements, and falls back to earth) requires some sort of guidance and control system.

The recent invention of the Global Positioning System (GPS), which is widely used in aircraft, ships, and automobiles, is a major tool for guiding a rocket. A brief description of the GPS follows.

Twenty-four satellites are continuously in orbit about the earth, about 12,000 miles up and making two orbits a day. Each satellite weighs about 2000 pounds and is powered by solar energy panels. This power is used to receive and transmit signals to the interrogating point. If this point can communicate simultaneously with at least three of the satellites, which is usually possible, then the longitude, latitude, and speed of motion of the interrogating point is determined. The location will be precise within ten feet. If the point can communicate with at least four satellites, then an additional piece of information, the altitude of the point, can be obtained.

The system works by the following principle. A pulse sent from the interrogating point is sent to a satellite and is promptly returned. The time elapsed gives a precise measure of the distance from the satellite to the interrogation point. The location of the satellite is known by a computer. From knowledge of the distances from three satellites with known locations, the precise location of the interrogating point is calculated by a computer.

Once the capability of launching satellites was achieved, in the early 1960s, this positioning system was rapidly developed by the military. In order for a submarine somewhere at sea, carrying nuclear missiles, to be able to fire a missile that can hit a distant target, the GPS is obviously useful. After a decade of use by the military, the system was made available for civilian use, anywhere in the world.

Let us discuss other means of controlling a rocket. First, consider an arrow shot from a bow. The arrowhead provides weight at the front end of the arrow, and the feathers at the back end are also important. The arrow does not tumble, but continues to fly with the arrowhead at the front. Arrows were developed long before aerodynamicists existed, but aerodynamicists can now explain exactly why the arrow behaves this way. Imagine that an arrow in flight starts to yaw slightly. Then the air resistance at the back, where the feathers (or fins) are, is greater than the air resistance at the front of the arrow. Thus, the force due to air resistance tends to correct the yaw, and the arrow continues its stable flight. It is also helpful that, because of the weight of the arrowhead, the center of gravity of the arrow is in front of what the aerodynamicist calls the center of pressure. Thus, stability is achieved.

Exactly the same principle applies to a rocket equipped with fins near the rear. As long as the rocket is flying through the earth's atmosphere, no tumbling will occur. Further, if the fins are movable, the rocket can be steered.

But, what do we do if the rocket is above the atmosphere (say, more than fifteen miles up)? Now, fins do not work. There are two regimes to consider; when the rocket motor is firing, and afterward, when the rocket is coasting through space.

When the rocket is firing, there are four possible ways we can take advantage of the propulsion to manipulate the thrust force. First, the rocket nozzle can be swiveled slightly sideways. The second possibility is to inject a fluid into one side of the nozzle. Thirdly, one can insert a heat-resistant body unsymmetrically into the exhaust jet. A fourth method can be used when the propulsion consists of a cluster of three or four liquid-propellant rocket motors, each with its own nozzle. Normally, the propellant would be fed at the same rate to each of the motors. When is it desired to steer the vehicle, the rate of propellant feed to one or two of the three or four motors can be reduced by a few percent. This "differential throttling" will provide steering.

What do we do after the main engines have shut down, and the spacecraft is coasting through the vacuum of space? The answer is to use small control rockets. If fired forward or backward, the craft is decelerated or accelerated. Or, if fired at right angles, a turning force is provided. If fired tangentially, spinning or despinning will be the result. These small control rockets are generally liquid-propellant rockets that would not use cryogenic fuel or oxidant. Various choices for propellants are available, and are discussed in a previous chapter.

One might ask, what is the role of the astronaut traveling in the spacecraft? As long as the radio communication between the spacecraft and a ground control point is working, guidance and control may be provided with no help from the astronaut.

9

MILITARY APPLICATIONS OF ROCKETS

Let us first discuss air-to-air missiles. These are usually solid-propellant rockets, launched from an airplane, and intended to destroy another airplane. Since the rocket is generally much faster than the target airplane, and carries an explosive warhead, the primary technical problems are guiding the missile to the target, and dealing with countermeasures the target airplane can take.

There are perhaps fifty different designs of air-to-air missiles, from a dozen countries. They vary in size, speed, and guidance. Let us describe a current United States missile, the AMRAAM. Launched from an airplane, it can fly 39 miles at Mach 4 (close to 3000 miles per hour). It weighs 335 pounds and has a 45-pound warhead. It is about 12 feet long and 7 inches in diameter. Its guidance involves mid-course updating and terminal active radar homing. See Figure 29. This and other types of guidance will be described next.

FIGURE 29. THE AMRAAM AIR-TO-AIR MISSILE

There are basically three types of guidance systems for air-to-air missiles, involving radar, infrared, and electro-optical imaging.

If radar is used, either the missile may carry its own radar transmitter, or the launch airplane may use its own more powerful radar to illuminate the target and maintain the target under illumination. The missile guides itself by detecting the radar reflected from the target.

There are three defenses available to the target. It may perform violent maneuvers. It may release chaff (small pieces of aluminum foil) that confuses the radar. Or, it may have sophisticated electronic countermeasures on board which detect that it is being illuminated by radar, and send back signals which confuse the radar.

Another guidance system uses infrared. The missile detects the heat waves from the target airplane, especially from the engine tailpipes, if the missile is approaching from the rear. Even if the missile is approaching from the front or side, the skin of the target airplane will be warmed by air friction, and a sufficiently sensitive infrared detector can pick this up. The airplane may be able to save itself if it launches one or more infrared-generating flares as decoys.

The third, most sophisticated, guidance system is electro-optical. The area in which the target may be located is scanned by an optical system on the missile. The optical image is converted to an electrical image. Once the target is seen, the missile is able to lock on to its image. Even if the target tries to defend itself by launching a flare, the missile can see that the target looks different from the flare, and it is not fooled.

Let us now discuss ground-to-air missiles. These are also intended to shoot down airplanes. They range from small enough to be carried and launched by a single person to quite large and launched from a trailer or a warship. They will often be two-stage solid rockets, because they have to climb to the target's altitude as well as catch it. Guidance is generally by radar or infrared, more or less as discussed for air-to-air missiles. Two examples will be described.

The STINGER, manufactured by the United States and used by many countries, is man-portable and infrared-homing. It weighs 33 pounds, with a 7-pound warhead. It is propelled by a two-stage rocket motor, and can reach speeds of 1700 miles per hour. Its range is about three miles.

The PATRIOT, also made in the United States, is much larger and more powerful. It is the primary surface-to-air missile used by the United States Army. Two separate trailers are used, one as the launcher, which may hold several missiles, and the other as a fire-control center, having a powerful radar. The missile is sixteen feet long and weighs 2000 pounds. It reaches a speed of close to 4000 miles per hour. See Figure 30.

FIGURE 30. THE PATRIOT SURFACE-TO-AIR MISSILE

The guidance of the Patriot is as follows. The ground-based radar locates and tracks the target, and guides the missile toward it. When the missile gets close, its own radar seeker detects radar reflection from the target, which is being "illuminated" by the ground-based radar.

The Patriot has also been used in tests to intercept an incoming ballistic missile.

The next category of missiles is air-to-surface. It has long been the practice for an airplane to drop bombs on a target below, and, relatively recently, smart bombs have been developed which have movable fins and, with appropriate guidance, can hit a target with great accuracy. The weakness of this approach is the vulnerability of the airplane, which must be essentially over the target. Attacking a surface target with a rocket-propelled missile has the advantage that the launching aircraft can be some distance from the target (standoff distance).

One example of such a missile is the HARM missile, made in the United States. It has a range of over 30 miles, and moves with a speed in excess of 1500 miles per hour, propelled by a solid-propellant motor. It weighs 807 pounds, and has a 146-pound warhead. It is guided by a passive radar seeker, which depends on reflection from the target of radar from the launch vehicle. It might be used to attack a warship.

Another example is the HELLFIRE missile, which can be launched from a helicopter, and is designed to attack tanks or other armored vehicles. The missile weighs only 99 pounds, with a payload of eighteen pounds. Its motor contains a double-base solid propellant chosen to produce minimum smoke, so as not to reveal the location of the launch vehicle. The launch vehicle "paints" the target with a laser beam and the missile homes in on the reflected laser light.

The final category of military missiles to be discussed is the surface-to-surface missile. The German V2 rocket, which introduced the modern era of rocketry, was a surface-to-surface missile (European continent to London). The STINGER missile, previously described as a surface-to-air missile, can also be used to attack a ground target. However, the most interesting surface-to-surface missiles are the intercontinental ballistic missiles (ICBMs). The first ICBMs developed by the United States were the Atlas and the Titan, both using liquid propellants, and the Minuteman, using solid propellants. We select the TRIDENT missile, shown in Figure 31, for detailed description.

FIGURE 31. THE TRIDENT INTERCONTINENTAL MISSILE

The Trident is a successor to the Polaris missile, first launched from a submarine in 1960. The Trident was put into service in 1990. The Trident is launched from under water, and has a range of over 7000 miles. It is launched from a tube by a gas generator, and once it has gotten safely away from the submarine, the first of its three solid-propellant rocket stages ignites. It weighs 130,000 pounds, and carries a payload of about 6000 pounds, which is large enough so that multiple nuclear bombs can be simultaneously released. The guidance system is interesting. It consists of inertial guidance, supplemented by a stellar sensor. Inertial guidance is self-contained, and is immune to jamming and deception.

Roughly, it works like this. A combination of gyroscopes and accelerometers can sense where the missile is at any moment, relative to its starting point. This information is fed to an on-board computer, which knows the location from which the missile was launched, as determined by GPS, and the target location. It calculates what corrections to provide to the trajectory during flight, and transmits this information to the missile. The stellar sensor provides additional accuracy.

The foregoing discussion gives examples of many types of military missiles. The majority of all rocket propulsion systems built today are for military purposes. The money invested by the military has been crucial to the rapid development of rockets since the V2.

There are a few additional uses of rockets by the military. Consider a cruise missile, which is launched from an airplane and flies a long distance at a low altitude (hopefully beneath the radar). For this application, an air-breathing jet engine is more appropriate than a rocket. However a solid-propellant rocket may be still needed as a booster to launch the cruise missile and bring it up to speed. Another military application is the use of rocket-assisted gun-launched projectiles for attaining longer artillery ranges. The small rocket motors attached to the projectiles are capable of withstanding very high accelerations in the gun barrel.

10

SPACE MISSIONS TO DATE

The very first space trip occurred on October 4, 1957, when the U.S.S.R. put SPUTNIK 1 into earth orbit. This spacecraft was an aluminum sphere weighing 183 pounds. It had antennas and was able to broadcast radio signals to earth. It completed about 1400 orbits (98 minutes per orbit) and then the orbit decayed because of air friction. If the orbits had been a little higher, Sputnik 1 would have stayed up much longer. It was launched from Baikonur Cosmodrome in Kazakhstan. Then, a month later, Sputnik 2 was launched. It weighed much more than Sputnik 1 and carried a dog. See Figure 32.

FIGURE 32. THE SPUTNIK

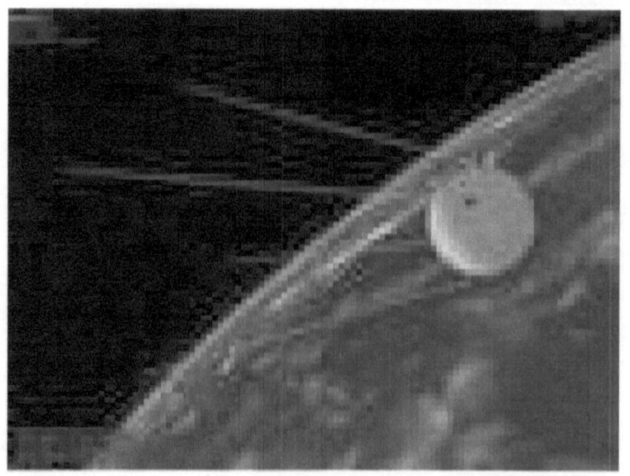

This was in the time of the Cold War between the United States and the U.S.S.R., and Americans were shocked that the Russians were ahead in what was called the space race. However, within four months, the United States was able to put something into orbit. On January 31, 1958, EXPLORER 1 was launched from Florida. It was assembled from existing rockets at Redstone Arsenal, Huntsville, Alabama by a team headed by Werner von Braun, who had been brought to the United States from Germany after World War II. In the following months, the United States formed the National Aeronautics and Space Administration (NASA) by merging units from the National Advisory Committee for Aeronautics and other government agencies. Also, at this time the U.S. Department of Defense formed the Defense Advanced Research Projects Agency (DARPA).

It is interesting to note that the U.S.S.R. rocket work traced back to Konstantin Tsiolkovsky and Sergei Korolev, while the U.S. rocket program can be traced back to Robert Goddard in Masachusetts and Werner von Braun in Germany, who was preceded by Hermann Oberth.

From 1958 on, many rockets were launched into space. The next major milepost was the entry of man into space. On April 12, 1961, Yuri Gagarin of Russia made the first orbit, in VOSTOK 1. This was a sphere 7.5 feet in diameter, weighing five tons. The vehicle only made one orbit, reaching a speed of 4.8 miles per second, and then it was decelerated by a retro rocket and re-entered earth's atmosphere. No provision had been made for a soft landing, and Gagarin safely ejected from the vehicle at 23,000 feet, as planned, and parachuted to the surface.

Ten months later, John Glenn became the first American to orbit the earth (MERCURY PROJECT). He made three orbits. See Figure 33.

FIGURE 33. THE MERCURY PROJECT

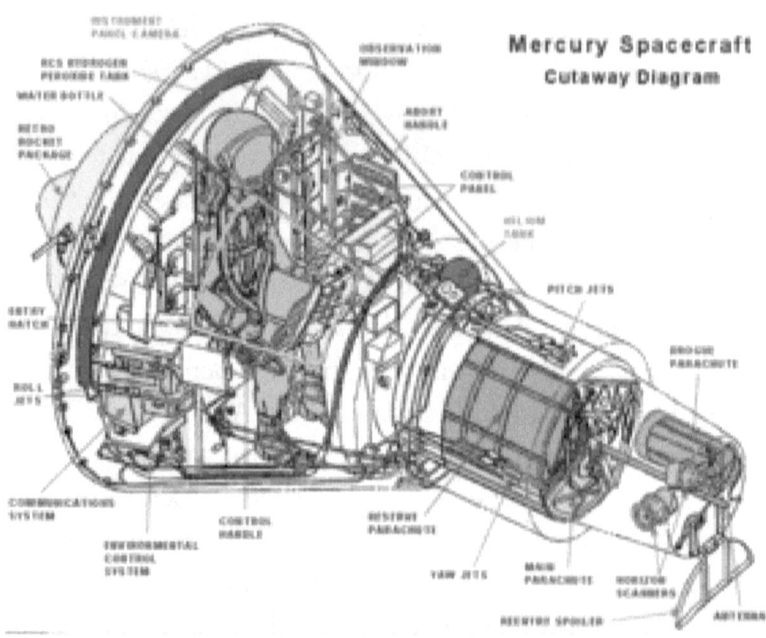

This led to a series of manned launches by both the United States and the U.S.S.R., culminating in the 1969 moon landing, to be discussed later. Meanwhile, in the 1960s, there was extremely active development of a communications satellite.

The concept was put forth by Arthur C. Clarke, a British writer best known for his science fiction, in 1945. His idea was to locate three satellites in geosynchronous orbit, 22,000 miles above the equator. At this altitude, they would orbit synchronously with the earth's rotation (that is, one orbit per day), and therefore each would always stay above the same point on the equator. Using power from solar collectors, they would relay messages, and even television broadcasts, from virtually any point on earth to any other point, except near the north or south pole. The transmitting stations would know exactly where the nearest satellite would be at any time, and they could aim their antennas accordingly. This would eliminate the need for cables under the ocean, and would permit communication with airplanes and ships at sea.

His concept was ignored until 1957, when Sputnik was launched. Then, the commercial advantages of Clarke's concept were widely recognized, and programs were initiated to put communication satellites into orbit.

The first commercial communication satellite, TELSTAR, was launched on July 10, 1962. This was a project supported jointly by AT&T, NASA, and British and French government agencies. It was not put in a synchronous orbit, 22,000 miles up, but in a much lower orbit, making one revolution every 2.6 hours. Accordingly, any transmitting or receiving station would be in contact with the satellite for intermittent, limited times. On the other hand, the satellite would be much closer to earth than a geosynchronous satellite, with less signal attenuation because of distance. The intermittency problem could be solved by having a large number of such satellites in low orbits, communicating with each other.

Then, many satellites were put into orbit, both with low orbits and geosynchronous orbits. (Improvements in receiver sensitivity took care of the distance problem.) In 1975 SATCOM 1 was put into synchronous orbit by RCA Americon and was the first satellite to broadcast ABC, NBC, and CBS television programs. Anyone in the world could watch Olympic sporting events at remote locations, or telephone anywhere in the world. Further improvements took place, and, currently, a satellite can handle hundreds of thousands of telephone conversations simultaneously. Furthermore, the communication satellites make global positioning systems (GPS) feasible. Such systems were discussed in the previous chapter.

Another notable event was the Gemini 7 flight in December 1965, when Frank Borman and James Lovell remained in orbit for 330 hours (almost two weeks).

Now we turn to the first moon landing, on July 20, 1969. This was accomplished by APOLLO 11, a United States program. The mission, and each of the five subsequent Apollo missions, involved three astronauts. In each case, a Saturn liquid-propelled rocket, developed at Huntsville, Alabama by Werner von Braun and his team, was used to launch a spacecraft carrying the astronauts and a lunar module. The spacecraft went into orbit around the moon after two days of travel time, and two of the three astronauts entered a lunar module that then descended to the surface of the moon and made a soft landing. The astronauts came out of the module and walked on the moon, planting a flag and collecting 48 pounds of rocks and soil, which they brought back. A television camera broadcast live pictures to earth of the astronauts on the moon. They spent 21 hours on the moon. The lunar module then ascended and made rendezvous with the spacecraft, and then they returned to earth. The mission lasted 195 hours. See Figure 34.

FIGURE 34. MAN ON THE MOON

The five subsequent NASA moon missions occurred in the next three and a half years. APOLLO 15, taking place in July 1971, was especially notable. David Scott and James Irwin stayed on the moon 67 hours, and explored the region near the landing site with a lunar rover. It was able to make excursions up to sixty miles away from the landing point. The last mission, APOLLO 17, in December 1972, was the last time a human has walked on the moon.

During the time the Apollo program was active, and afterward, the Russians have carried out many space missions. The third country to enter the age of manned space exploration is China, who in October 2003 launched SHENZHOU 5, which carried astronaut Liwei Yang around the earth 14 times.

In the space age to date, hundreds of missions have been made. A few of them ended in disaster for the crew. Many of the missions since the Apollo program have been unmanned, and have used sophisticated instruments to obtain a variety of scientific data, including close-up pictures and spectrograms of planets, moons of planets, and comets, and even chemical analysis of the materials on the moon and Mars. Some missions have concentrated on the earth, obtaining data of meteorological and agricultural value, and accurate measurements of the earth's gravitational field and magnetic field. Some missions have delivered astronomical instruments to space, so that the universe can be observed above the earth's atmosphere. Some have placed communication satellites and "spy" satellites in orbit. Many have been devoted to setting up and manning the International Space Station. Most remarkably, robotic vehicles have been transported to Mars; they are able to take and analyze soil samples and radio back the results. Some of these developments will be discussed in the following pages. See figure 35.

FIGURE 35. MARS, FROM ROVER

Something should be said about the disasters in the space program, alluded to above. It is obvious that space travel is risky. There were three major accidents occurring in the American programs. The first major accident occurred early in the Apollo program, on January 27, 1967. The accident occurred during a launch simulation procedure. Three astronauts (Gus Grissom, Roger Chaffee, and Ed White) were in the capsule, in a pure oxygen atmosphere, when a fire broke out, with fatal results. The fire was thought to have started because of an electric arc. They tried to escape by opening the hatch, but it was designed to open inward, and the increase in pressure in the capsule caused by the fire made it impossible to open the hatch. After the accident, it was decided to use a 60-40 oxygen-nitrogen mixture instead of pure oxygen during the launch period, and to change the design of the port so that it was easy to open and would swing outward. The Apollo program continued until after Apollo 17, and there were no more fatal accidents in that program.

The second fatal accident occurred on January 28, 1986 (nineteen years later). One of the Space Shuttle vehicles, named Challenger, exploded 73 seconds after launch, killing all seven astronauts on board. (The Space Shuttle is described in Chapter 4.) There had been eight previous Challenger flights, with no serious problems. The accident investigation found that a rubber "O-ring" sealing a joint in a solid-propellant rocket booster failed, causing flame to shoot out of the opening and impinge on the hydrogen-oxygen main engine, causing the explosion. It was thought that the rubber "O-ring" had become brittle in the hours before launch, when the ambient temperature was unusually low for Florida. Also, the "O-ring" design was judged to be faulty. A new design was used on future Space Shuttle boosters, and this problem did not recur.

The third serious accident occurred on February 3, 2003 during return to earth of Space Shuttle Columbia. This was 17 years after the previous Shuttle accident, with many successful flights in between. There were seven astronauts aboard. All died. The problem rose during re-entry into the atmosphere. When a spacecraft (or meteorite) encounters the atmosphere at very high speed, friction causes the outer surface to become very hot. To avoid damage to the spacecraft, ceramic tiles and reinforced carbon panels are provided to protect critical parts. On this mission, apparently, some of these protective elements were dislodged during the launch. During the mission, the astronauts were unaware of the problem. Upon re-entry, about 16 minutes before the scheduled landing, the heat fatally damaged the spacecraft. In flights after this one, means were found to inspect the protective elements while in orbit, and to repair or replace elements if necessary.

The Russian space program also suffered some fatal accidents, but complete details are not available.

In spite of these accidents, keep in mind that hundreds of successful space missions were accomplished in the same time period.

Let us return to discussion of some of the successes of the space program. The placing in orbit of telescopes, spectroscopes, cosmic ray detectors, and many other instruments have led to much new knowledge about the universe. Special mention must be made of the Hubble Space Telescope. It was put into orbit at an altitude of 360 miles in 1990 and has been enhanced since by five servicing missions. It weighs twelve tons, is 43 feet long, has 384 square feet of solar arrays for power, has a primary mirror 8 feet in diameter, and a spectral range from ultraviolet to near-infrared. It can be maneuvered from earth by radio. Being above the earth's atmosphere gives it an enormous advantage over earth-based telescopes. See Figure 36.

FIGURE 36. THE HUBBLE TELESCOPE

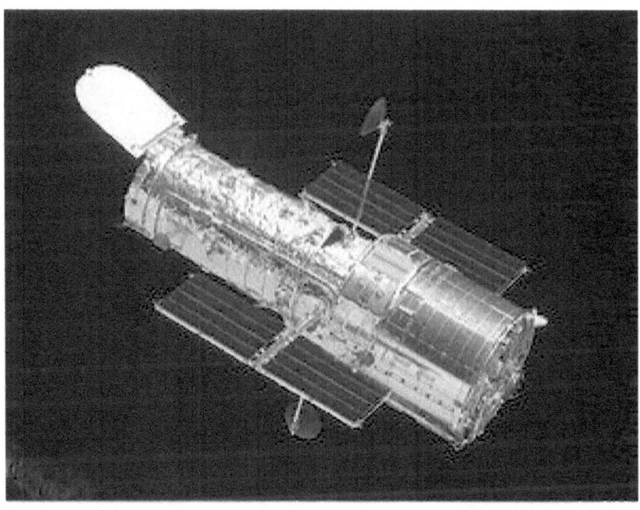

It gives unprecedented clear views of the universe. It can observe far more distant galaxies than any earth-based telescope, and this has enabled astronomers to deduce that the universe is much older than previously thought; namely fourteen billion years old. It has provided evidence for "dark matter" and "dark energy" in the young universe. It has discovered a galaxy 2.6 million light-years away which is the most powerful known source of radio waves. It has observed two galaxies in collision, with formation of new stars. It also provides superior looks at nearby planets, moons, and comets.

For astronomers, this instrument alone provides a complete justification for the space program. However, it could be argued that communication satellites represent a more important practical benefit. Also, observations of Earth from above are providing key information about weather patterns, crop growth, and evidence of global warming shown by polar ice-pack reduction.

Another major project is the International Space Station, shown in Figure 37. The Space Shuttle, previously described, is the principal ferry to it from earth, although Russian rockets have also ferried people and supplies to it. Construction of the Space Station was commenced in November 1998, with a Russian spacecraft serving as the first component. Additional components have been added from time to time, and completion is estimated to occur in 2011. It is jointly supported by five space agencies. At the present time (2010), its current weight is 759,000 pounds; including solar arrays its width is 240 feet, its length is 171 feet, and its height is 90 feet. Additional components will be added. It is designed for a crew of six, and it has been continuously manned since October 2000. The permanent crew is from the United States or Russia, but it has been visited by astronauts from 15 countries. It makes 16 earth orbits a day at an altitude varying from 173 to 286 miles, and a speed of 17,000 miles per hour

FIGURE 37. THE INTERNATIONAL SPACE STATION

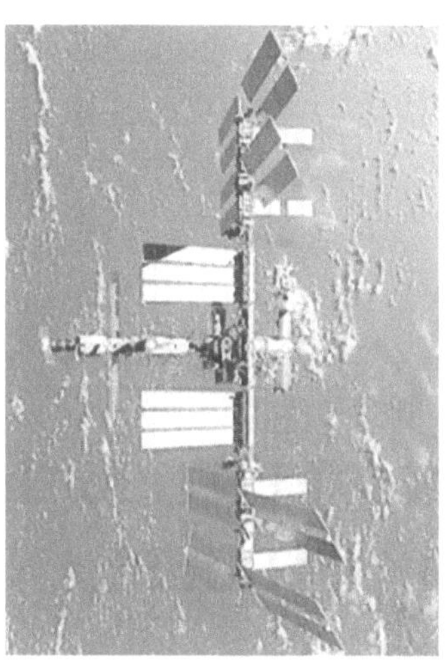

What is the purpose of the Space Station? There are three possible uses. First, it provides a way of determining how well a human can survive in space. This is a necessary preliminary to a long manned space flight, say to Mars. To date, three astronauts have each spent over a year in space. Second, it provides a laboratory in which experiments can be done in very low gravity, virtually zero gravity, and above the earth's atmosphere. Examples are biological experiments, meteorological and astronomical observations, crystal-growing experiments, and combustion experiments. (Will a candle continue to burn in the absence of a gravitational field, which causes convection of air into the flame? Answer: yes, but differently.)

Third, the Station could be used as a platform for launching spacecraft into deep space. Since the Station is already moving at nearly five miles per second, a vehicle leaving it would only need a modest velocity increase to escape the earth entirely. However it has been argued that the Space Station is an unnecessary complication, since a vehicle bound for Mars can leave directly from the earth with no intermediate stop, and in fact has already done so.

Obviously, the experience gained in operating the Space Station will be very useful for the time when manned visits to distant parts of the solar system will occur. This must be balanced against the high cost of shuttling supplies from earth to the shuttle, presently estimated to be thousands of dollars per pound.

11

TRAVEL TO MARS AND OTHER PLANETS

Rockets provide the means for travel to Mars. Why is travel to Mars of such interest? The first reason usually given is to search for signs of extraterrestrial life. An early stimulus was the claim made in 1877 by Italian astronomer Giovanni Schiaparelli that he could see straight lines on the surface of Mars with his telescope, which he called "canali", or channels. This was mistakenly translated into English as canals, and people assumed that they were artificial waterways, which would be strong evidence for intelligent life on Mars. Other astronomers claimed that they, too, could just barely see these canals. In 1898 the English writer H. G. Wells wrote a popular novel, "The War of the Worlds", describing an invasion of the earth by Martians. Wells was undoubtedly influenced by the reports of canals on Mars.

However, once it was possible to send a rocket on a flyby to Mars and transmit high-resolution pictures to earth (1965), it was clear that there were no canals in Mars. Obviously, astronomers straining to see Mars through the earth's atmosphere, which distorted vision, had good imaginations but were mistaken. Figure 38 shows a close-up picture of Mars from a Mariner spacecraft.

FIGURE 38. MARS, AS SEEN FROM MARINER

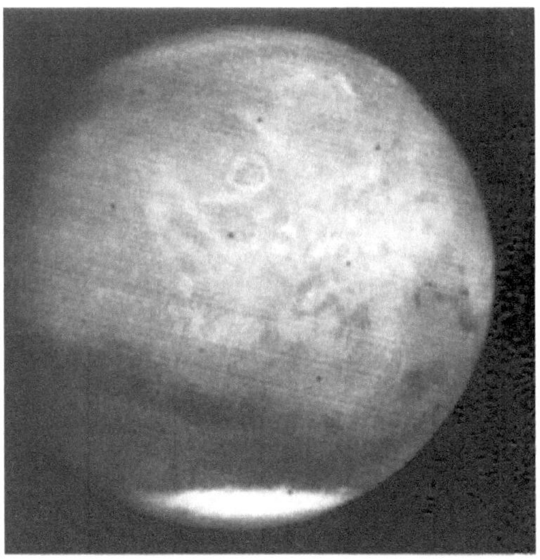

Nevertheless, the possibility exists that there is or at one time was life of some form on Mars. Why would this be so important?

What do we know about life? Biologists have discovered that all forms of life on earth, plants and animals, have cells containing DNA, a complex organic molecule that differs significantly for each species, but has elements in common in all species. This strongly suggests that all life evolved from an initial living cell.

There is good evidence that the earth is four and a half billion years old. Until three billion years ago, it is believed that conditions on the earth's surface were too severe for any form of life to survive. Then, mysteriously, the first living cell that could reproduce itself appeared. In the next two billion years, primitive unicellular life forms developed the ability to obtain energy from sunlight by the process of photosynthesis. In photosynthesis, oxygen is produced. Earth's original atmosphere is believed to have been oxygen-free, but photosynthesis caused oxygen to build up to the present 21 percent.

As of a billion years ago, plants but no animals existed. Plants have very low need for energy but animals require more energy, for mobility. Animals get their energy by oxidizing organic compounds, so they cannot exist unless oxygen is available. In the past billion years, plants and animals evolved to what we have today.

There is extensive evidence for the foregoing sequence of events, largely from fossils, but scientists today lack understanding of how the first living cell appeared, even though they now have good understanding of the reproduction process, involving DNA. If life can be found on any other planet, the key question is, does it have DNA, and, if not, how does it reproduce itself? Such information would be important in addressing the mystery of how life first appeared on earth. Greater knowledge of life could lead to cures of various diseases.

The main facts about Mars are these. It is the fourth planet from the sun, averaging 140,000,000 miles from the sun. Earth is about 93,000,000 miles from the sun, so sunlight reaching Mars is only 43 percent as intense as sunlight reaching Earth. The sunlight reaching the two closest planets to the sun, Mercury and Venus, is much more intense than what reaches Earth, and both these planets are far too hot for liquid water to exist on their surfaces.The four most remote planets, Jupiter, Saturn, Uranus, and Neptune, get so little sunlight that they are far too cold for liquid water on their surfaces, unless heat is available from some subsurface source (volcanic activity?).

It seems very likely that liquid water is needed as a precursor to life. If this is so, then the only planet with a good chance to harbor life, other than our own, is Mars. Hence a strong interest in traveling there. (A moon of Saturn named Titan is thought to be another possible place while life might exist.)

We know that Mars' diameter is only 53 percent of Earth's, and the strength of gravity on Mars' surface is 38 percent of the Earth's 32.2 feet per second per second. Accordingly, a rocket can escape from Mars more easily than from Earth. We also know that Mars has an atmosphere whose pressure at the surface is less than one percent of ours. Its thin atmosphere consists mainly of carbon dioxide. Its mean surface temperature is minus 81 degrees Fahrenheit.

To date, there have been at least fourteen successful one-way trips to Mars, the first occurring in 1965. Of these, three were launched by the U.S.S.R., one by the European Space Agency, and the rest by the United States. Rather than providing details of each mission, only selected highlights will be presented.

Mariner 4 (1965) made the first successful flyby, sending back 21 pictures of Mars.

Mars 3 (1971), launched by Russia, involved an orbiter and a lander. The lander was only able to transmit data for 20 seconds after landing, but the orbiter continued to send back data for 8 months.

Viking 1 and Viking 2 (1975), each an orbiter and a lander, made successful landings and sent back data on the Martian atmosphere and results of soil experiments.

Odyssey (2001) sent back many high-resolution pictures, and was still functioning in orbit five years later. It acts as a communication link between the two Rovers (below) and Earth.

Mars Exploration Rovers Spirit and Reconnaissance (2003) both made successful landings and, between them, sent back more than 100,000 pictures. They were able to move for miles over the Martian surface. Their ability to function lasted far longer than expected.

None of these expeditions have sent back good evidence of life in Mars. There is evidence of water on Mars, of four types. Hematite deposits on the surface of low areas have been seen that may have been deposited in the past when water evaporated. Ice is present at the poles, according to spectroscopic observations by orbiters. Gullies seen from above are taken as evidence of erosion by water. Recent changes in these gullies suggest that water periodically emerges from underground. Finally, recent observations have shown that there is a very small amount of methane gas in the Martian atmosphere, which could have been generated from once-living matter. Alternatively, it could have come from volcanic eruptions, but there is no sign of active volcanos on Mars. Further, volcanic gases on earth always contain sulfur compounds, and there is no trace of such in the Martian atmosphere. Another explanation for the presence of methane is that it results from reaction of certain minerals with water, at elevated temperature, to form hydrogen, and such hydrogen could conceivably react with carbon or carbon dioxide to form methane. This is highly speculative.

On reviewing these accomplishments, we see a sequence from flybys to orbiting to landing. We see that there is no problem in communicating. The next major goal is to send an unmanned vehicle to Mars and return it to Earth, with samples. After that, a round trip involving a manned landing would be the next goal, but that is probably decades in the future.

Future trips to Mars, either manned or unmanned, may well be able to prove that water exists there and may find more positive evidence for past life there.

Spacecraft have been sent to planets other than Mars. About twenty spacecraft have been sent to Venus by NASA, by the USSR, and by the European Space Agency.

The first successful probe into the Venusian atmosphere was in 1962, and the first landing in 1966. The landing probe only survived briefly,

It should be noted at once that Venus is very hot, with a surface temperature of 900 degrees Fahrenheit. Its atmosphere consists mainly of carbon dioxide, with clouds of sulfuric acid droplets. The atmospheric pressure at its surface is ninety times that of the earth. No form of life as we know it could exist under such conditions. While the surface cannot be seen, subsequent missions using radar have shown the presence of mountains, volcanoes, and meteor craters.

The Galileo probe made flybys of the four moons of Jupiter in 2003, and then descended into its atmosphere, being destroyed thereby. Jupiter is a gas giant (no solid surface). The probe discovered ammonia clouds and numerous lightning strikes. One of the four moons was found to have a magnetic field.

NASA's Cassini spacecraft, carrying a Huygens probe built by the European Space Agency, has explored Saturn and its moon Titan in 2005 and thereafter. While Saturn is a gas giant, Titan is solid, and is interesting. Its surface temperature is minus 288 degrees Fahrenheit, and lakes of what appear to be liquid methane are on its surface. If there is a subterranean source of heat, life in some form might be able to exist.

12

POSSIBLE FUTURE DEVELOPMENTS

There are three categories of goals. First, there are those goals that clearly can be achieved if substantial effort is made. Second, there are goals that are more difficult, because it is far from obvious how to proceed, but even so we know of no scientific fact that makes the goal impossible. Third, there are goals that are impossible to achieve, based on current knowledge, but we can always hope that some new scientific discovery will change the situation.

Consider the question "Can man escape from the earth, and visit the moon and planets?" If this question were posed in the year 1880, for example, the scientists of the day would probably say that we know of no reason the goal is absolutely impossible to achieve, but we do not know how to proceed to achieve it. This puts the question in the second of the categories, above. However, if this same question were posed in the year 1910, after Konstantin Tsiolkovsky had figured out how the concept of staging and the use of liquid-propellant rockets would make the task feasible, optimistic scientists would have to say that it could be achieved in the next century. It turns out that it took 60 years. It could have been reached sooner if more concentrated effort had been made. Thus, after 1910. the question would fall in the first of the categories above.

Now, consider the question "Will man ever be able to visit planets outside our solar system, with a travel time less than his lifetime?" We can make an argument that such a goal falls into our third category, that is, impossible according to current scientific knowledge. However, history is replete with examples of revolutionary discoveries that drastically modified our views of what is possible. The discovery of radioactivity is one such example. Another is the discovery that the earth's continents drift from one location to another. Again, there is the discovery that diseases are caused by germs. What about the discovery of long-distance radio communication? Perhaps the discovery that the sun and not the earth is the center of our solar system should be listed. Hence, a future radical discovery may be made some day that would make manned travel outside the solar system possible.

Each of the possible future developments discussed in the following text falls into one of the three above categories. Travel outside the solar system falls in category three. Drastic improvement in the launch problem falls in category two. The other goals fall in category one.

12-1. SUPERSONIC COMMERCIAL AIR TRAVEL

The Concorde and the Tupolev TU 144 have already demonstrated the possibility of commercial air travel at Mach 2, but improvements are needed to make this economically viable. A question is, what about travel at much higher speeds, so that a trip from New York to Tokyo or London to Australia can be accomplished in a few hours?

Current development of commercial jets has been helped enormously by previous military research. Indeed, the development of the turbojet engine occurred just before World War II, with the thought that a faster military plane would result. Also, note that the commercial Boeing 707 is closely modeled after previous bombers. However, there does not seem to be a need for military transportation at Mach 10. The armed services have long had supersonic bombers. So, where will the development money for commercial supersonic planes come from? There are outstanding problems of high fuel consumption, low passenger capacity, and sonic boom. (At one time it was thought that high-flying planes would have an adverse effect on the earth's protective ozone layer, but it is now believed that this is not a problem.) Research has shown that better shaping of the fuselage can reduce the sonic boom by a factor of two.

In 1966 the United States government awarded a contract to Boeing to develop a supersonic transport better than the Concorde. Boeing designed for Mach 2.7 (compared with Concorde's Mach 2), 250 passengers (compared with Concorde's 100), and a range of 4000 miles. It would fly at above 60,000 feet. It would be built largely of titanium (compared with Concorde's aluminum). However this program was cancelled in 1971, largely because of Congressional concern at that time about possible destruction of the ozone layer.

In 1994 the European Supersonic Research Program was started, with joint support by the British, French, and Germans. The goal was to develop a second-generation Concorde, to enter service by 2010. It would fly at Mach 2, carry 250 passengers, and have a range of 6300 miles. This program seems to have been replaced by a four-year project launched in May 2005 with 37 partners from 13 countries, again seeking an acceptable design for a commercial supersonic plane,

In 2005, a Japanese-French joint venture was started to do research on a supersonic plane to fly by 2015. In 2006, a joint venture by Alenia, an Italian company, and Sukhoi, a Russian company, was announced to develop a supersonic business jet.

Both NASA and the Boeing Company are giving some thought to developing a supersonic commercial plane, but this does not seem to be a major priority of either of these organizations at this time.

12-2. TRAVEL OUTSIDE THE SOLAR SYSTEM

Why would scientists today say that manned travel to a planet outside the solar system is impossible? How far would one have to go to reach a planet outside our solar system? In the past few years, astronomers have discovered several hundred planets rotating about distant stars. If we were planning a journey, we would seek a planet as near as possible, with a size and temperature close to that of Earth, to maximize the chance that some form of life has developed on it. Our best telescopes are not able give us much detail about these distant planets, so far. In April 2007, astronomers announced that they had found a good candidate. This planet,

20 light-years away, orbits about a dim red star, Gliese 581. It is about five times as massive as Earth and orbits its sun every thirteen days at a distance of seven million miles. While this planet is close to its sun, relative to Earth's sun, it's sun radiates only about one percent as intensely as our sun. Astronomers estimate that its surface temperature is near 20 degrees Celsius. Accordingly, it seems possible that life could exist on this planet.

If we could travel with the velocity of light, we would reach this star and its planet in 20 years. However, Einstein has proven that infinite energy would have to be expended to accelerate a space ship to the velocity of light. Suppose we settled for one-fourth of the velocity of light. At this speed, it would take 80 years to reach this planet, which would require a long-lived astronaut, or several generations of astronauts. We can calculate the minimum amount of energy that would have to be expended to accelerate a 3000-pound space ship to one-fourth of the velocity of light.

When we perform this calculation, neglecting the weight of the power plant (whatever it may be), we find that a trillion (that is, a million million) kilowatt-hours of energy are needed. How could such a quantity of energy be generated or stored in a 3000-pound spaceship? Three possibilities to consider are nuclear energy, solar energy, and energy generated on earth and beamed to the spacecraft by microwaves.

There are about 100 nuclear reactors in the United States, together capable of generating about 800 billion kilowatt-hours of electricity per year. If all these nuclear reactors were put aboard the 3000-pound spacecraft, then, working together, they would take over a year to generate a trillion kilowatt-hours. Clearly, the array of nuclear reactors would not be weightless, but would weigh far, far in excess of 3000 pounds. Therefore, we conclude that nuclear energy as we know it is not the answer.

If we make a similar analysis of solar energy or microwave-beamed energy, again we find that there is no way we can produce a trillion kilowatt-hours in a 3000-pound spacecraft in a reasonable time, because of weight considerations.

However, there is another way to analyze the situation. Assume that unlimited energy is freely available on the vehicle, with no weight penalty.

How would this energy be used? In order to accelerate the vehicle forward, it is necessary to expel a mass of propellant backward, preferably at very high speed. Unless the specific impulse is very, very high, an incredible amount of propellant would be required. (The higher the specific impulse of a rocket, the less propellant is needed.) Let us assume a specific impulse of 500,000 seconds. This would correspond to the exhaust leaving at 3000 miles per second. This about 100 times as fast as has been achieved in advanced experimental electric propulsion thrustors, and about 1200 times as fast as the exhaust from a hydrogen-oxygen rocket. Nonetheless, using this high specific impulse, we calculate that 20,000,000 pounds of propellant would be needed to accelerate a 3000-pound craft to one-fourth the velocity of light. If we assumed any lower value of specific impulse than 500,000, the mass of propellant would even be greater.

Summarizing, we have to conclude that, within the scope of present knowledge, there is no possible way to get a manned spacecraft to a planet 20 light-years away, in a man's lifetime, or even in several generations. Accordingly, no great effort is being expended on thinking about manned interstellar travel, except by science fiction writers. If some fantastic scientific breakthrough would occur, the picture could change.

There is still the possibility of sending a small *unmanned* spacecraft to a planet outside the solar system. Current developments in nanotechnology suggest that such a spacecraft could be exceedingly small, with a correspondingly reduced propulsion energy requirement. However, to be useful, such a spacecraft would have to possess the capability of telemetering information back to earth.

If we were to confine our space travel plans to our own solar system, what further technological progress is needed? We have already sent unmanned instrumented probes to virtually all the planets, and we have sent men to the moon a half dozen times. What problems remain to be solved?

We would like a more cost-effective way of leaving the earth than by a giant rocket like the Space Shuttle. We would like to able to reduce the travel time to a distant planet in our solar system. We need to know more about how long a person can survive in space, in view of the threats of cosmic rays and meteorites, as well as the long-term effect of low gravity.

We need to find out if a permanent space colony on the moon or Mars is feasible. And of course the astronomers have many more questions about the universe, which will require the design and orbiting of specialized instruments. These topics will be discussed.

12-3. THE LAUNCH PROBLEM

The present Space Shuttle is scheduled to be retired soon. NASA is currently designing the next generation, which will not only have the capability of going into orbit at less cost than with the Space Shuttle, but also will be able to go directly to the moon or Mars. At least, it is hoped that this is achievable. The design for this spacecraft, a so-called SPACE PLANE at one time said to be called Orion, has not yet been finalized. Cost saving would be accomplished by the reusability of the craft. The first manned test flight of Orion is not yet scheduled.

One concept of this vehicle is to mount it on a rocket, which would be the first stage, and then it would use its own power plant, acting as a second stage, to go into orbit. On returning, after re-entry it could fly to a landing place and then be reusable.

Another concept is to launch the space plane from a fast airplane flying east, and then it would accelerate with its air-breathing engine, perhaps a ramjet. After it reached its maximum speed with this type of propulsion, its rocket engine would be used to reach orbit. Possibly something more than one-stage rocketry would be needed to reach escape velocity.

It is too early to estimate how much more cost-effective the Orion will be than the present Space Shuttle.

Looking further into the future, we see that some significantly better propulsion system than hydrogen-oxygen, if found, will provide major improvement in launching. All imaginable chemical propulsion systems have been considered for the past fifty years, and no practical candidate better than hydrogen-oxygen has emerged.

Some study of a NUCLEAR ROCKET has been carried out. The idea would be to heat hydrogen to the highest possible temperature without meltdown by means of a nuclear power plant, and then expand it through a rocket nozzle. Ground tests of such a system have shown a specific impulse of a little more than twice that of hydrogen-oxygen. There are some strong negatives. If an accident occurred during launch, radioactive material would be strewn about. Aside from this, the reactor, plus shielding necessary to protect the crew, would probably be far too heavy.

There seems to be little prospect of dramatically improving the launch situation by rocketry, at present. However the far-fetched idea of a SPACE ELEVATOR has been proposed. Suppose a geosynchronous space station, remaining stationary 22,000 miles above a spot on the equator, could be connected to earth with a cable. If so, a "climber" going up and down the cable could carry up packages that could be assembled and then easily be launched into space. Note that the space station at the top of the cable is moving at 1.89 miles per second because of the earth's rotation, and the earth's gravitational attraction is much weaker than at the earth's surface. Therefore it would be relatively easy to launch a spacecraft away from earth in an eastward direction from such a space station.

This strange idea was first proposed by Konstantin Tsiolkovsky in 1895. The Eiffel tower was being erected at that time, and Tsiolkovsky had the idea of building a tower 22,000 miles high, containing an elevator. More thought by others led to the idea that you didn't need a tower; all you needed was a cable, connected to an orbiting body at the top. The orbiting body would always stay above the same spot.

The question arose, how strong would such a cable have to be? Unless it was very strong, the weight of the cable would tear itself apart. A steel cable would have only one-fifteenth the strength needed. Consideration is being given to a diamond filament 22,000 miles long, but no one knows how to make even a short diamond filament. Another possibility is to make a cable of carbon nanotubes, which might have the required strength. Again, this is far beyond the state of the art. Maybe in a hundred years a suitable cable material will appear. Meanwhile, it seems we are stuck with current chemical propellant technology.

12-4. LOW-THRUST PROPULSION

Even if we can't make a drastic improvement in the launch situation, there is another possible application of an advanced propulsion system. At present, a round trip to Mars would take at least two years, and if we wanted to make a round trip to the outer planets, much more time would be needed. Suppose we were to launch a rocket to a distant planet by conventional means, so it would begin its journey at a few miles per second, but we had another propulsion system aboard which was capable of producing a very small thrust force for a very long time. If it took a month or two to accelerate to a significantly higher speed, this would be worthwhile.

For example let's assume that the velocity after launch is four miles per second, and then a low-thrust propulsive engine is started. Assume that the engine generates one pound of thrust continuously for two months. Assume that the weight of the vehicle is 3000 pounds. If this engine can accelerate the vehicle from four miles per second to eight miles per second, this would cut the travel time of a long journey in half.

This would be easily achievable if a small engine using storable liquid propellants is used (provided the rocket nozzle survives for two months), except for the problem of the weight of propellants consumed. For example, suppose a hydrazine-nitrogen tetroxide engine was used, with a specific impulse of 280. A calculation then shows that 28,000 pounds of propellant would have to be carried on this 3000-pound spacecraft. Thus, it wouldn't work.

However, if we had a propellant with a specific impulse of 3500. this would change the picture. We would only need 420 pounds of propellant, which is not impossible for a 3000-pound spacecraft. But, how could we achieve such a specific impulse?

The only imaginable way is to use electricity to accelerate a working fluid to very high velocity. This can be done, at least on a small scale, using either electrostatic or electromagnetic thrusters. An electrostatic thruster has been built, operating with specific impulse of 3500, but the thrust is only one-twentieth of an ounce. It uses xenon as the working fluid.

The xenon is ionized and then electrostatically accelerated to very high speed. More research and development is needed to develop a large enough thruster of this type. Possibly even higher specific impulses than 3500 can be achieved.

There is still the need for a source of electricity, which could either be solar or nuclear. The previous discussion of power for interstellar travel still applies, but the problem is much easier because we don't need a trillion kilowatt-hours of energy; we can calculate that only 24,000 kilowatt-hours is required. Solar cells to deliver this much energy over a period of two months would weigh about 370 pounds and extend over an area of about 800 square feet, if currently available solar cells were used. (This is based on a solar cell being able to produce 45 watts per pound of weight and about 20 watts per square foot.)

Then, based on current solar cell technology, our postulated 3000-pound vehicle would need 420 pounds of xenon propellant and 370 pounds of solar cells.

If the weight of the accelerating device were no more than 200 pounds, the vehicle would still have 2000 pounds available for structure, passengers, and life support. Possible future improvements in technology may yield lighter solar cells and more powerful thrustors, resulting in even a more favorable situation.

We conclude that it is entirely within the range of possibility to add several miles per second to a space vehicle by use of a small thruster operating a long time, substantially shortening the travel time to Mars or outer planets. Some engineering development is needed.

12-5 LONG-TERM SURVIVAL IN SPACE

Up to now, some astronauts have spent up to one year in space, and have survived. The main problems are the lack of gravity, which causes muscles to atrophy, and radiation, which is not experienced on earth because of the protective effect of the atmosphere. The earth's magnetic field also provides protection from radiation near earth but not in deep space. For an astronaut outside the spacecraft (extravehicular activity), meteorites pose a hazard. There may also be a psychological problem caused by individuals confined in a small space for a long time. Continuing research on the International Space Station is expected to provide more information, and may lead to ways of improving the situation. However, the Space Station is not high enough to be above the earth's magnetic field to permit study of what would happen to an astronaut on a space walk far from earth's magnetic field.

12-6. PERMANENT COLONY ON THE MOON

NASA has announced an objective to establish a permanent colony on the moon by the year 2024. It is planned to locate it near the moon's south pole, so as to have the maximum access to sunlight for solar cell operation. It is hoped that other nations will participate with the United States in this project. There may be commercial as well as scientific activities in the colony. The plan includes having a pressurized roving vehicle that would take people on expeditions far from the base.

It is conceivable that rocket fuels could be manufactured from indigenous materials, especially if water can be found. Possibly valuable minerals will be discovered. Certainly wealthy tourists might want to visit a moon colony. Astronomers could erect very large telescopes, because of the lower gravity, which might see more than the Hubble telescope.

There is no major technical problem to prevent establishing a permanent moon colony, once a suitable rocket-propelled lunar lander is designed and built. The main uncertainty about a moon colony is associated with financial support.

12.7 UNMANNED PROBES

There are many scientists who believe that there is no good scientific reason to send humans into space, as long as unmanned probes can be used for exploration. The probes can be much smaller and less costly. A counter-argument to this is that humans can go into space, and have already done so, to repair the Hubble telescope and various earth satellites.

There is also a fear that if the United States loses the ability to send men into space, some other nation with that capability could create problems, for example disabling our observation satellites.

Perhaps a distinction can be made between human space travel near the earth and human travel to more distant points.

12-8. ASTRONOMICAL GOALS

Probably the most interesting question has to do with the origin of the universe. Did it start with a "big bang"? Today scientists say yes (99 percent certainty), because of microwaves coming from all directions that can only be explained as a residue of the big bang, coupled with the recession of the galaxies from one another as discovered by Hubble. How long ago did this occur? Currently, based on observed movements of galaxies, we think it happened 14 billion years ago. What happened during this time span? How did gases condense into stars? What are "black holes"? What is "dark matter"? What is "dark energy"? Where do cosmic rays come from? While astronomers have tentative answers to some of these questions, there is much uncertainty, and better astronomical observations will probably lead to clearing up uncertainty. These are the most important questions in cosmology.

Let us consider "dark matter" for a moment. It cannot be seen, but its presence can be inferred from observed gravitational effects on stars and galaxies. The astronomers have calculated that the vast majority of mass in the universe consists of "dark matter". Currently, no one knows what this dark matter consists of. One theory postulates that it consists of a large number of black holes. Another theory assumes that it consists of a new type of elementary particle which does not readily interact with ordinary matter, and hence has not been detected. Cosmic rays might be generated by high-velocity collisions of such particles. It is possible that currently planned new instruments for measuring cosmic rays in space, once operating in orbit, may provide some clues. Future understanding of dark matter or dark energy could conceivably lead to new energy sources on earth.

Another fascinating question has to do with life in the universe, other than on earth. If it exists, will it have DNA? If so, how does that DNA differ from our DNA? To date, no evidence has emerged that life exists anywhere else. However, we know that trillions of stars exist, and recent observations have shown that planets revolve around many of the closer ones. We presume that there could be billions of planets in the universe that are similar to the earth in temperature and chemical composition. We have not detected radio signals indicating intelligence from any of them, to date. So far, we have found no conclusive evidence of life on other planets in our solar system, but it is possible that planned trips to Mars or a moon of Saturn will find something. As previously discussed, we do not know how to travel outside the solar system, but we can explore by radio, and have been doing so for some time. A program called SETI (search for extraterrestrial intelligence), which has been active for several decades, is now using more advanced detection equipment to scan the skies for incoming radio signals of a very large range of frequencies. I am sure you will hear if success occurs.

The absence of radio signals to date suggests the absence of intelligent life, but even the discovery of unintelligent life would be of enormous importance to biologists, as well as theologians.

APPENDIX A

NEWTON'S LAWS AS APPLIED TO JETS AND ROCKETS

Newton's First Law of Motion states that a body at rest will remain at rest unless acted upon by an unbalanced external force. A body in motion at a given velocity will continue to move at that velocity unless acted upon by an unbalanced external force.

Applying this to a turbojet-propelled plane flying at a constant velocity, the law implies that it will continue to fly at that velocity without requiring thrust, unless acted upon by an unbalanced external force. However, there are two forces to consider, gravitation and aerodynamic drag. The gravitation force is balanced by the lift provided by the wings. However the aerodynamic drag (friction of the air) is an unbalanced force, and the thrust of the engine is needed to balance this force so that the airplane can maintain constant velocity.

Applying this law to a rocket, once again the two forces to consider are gravitation and aerodynamic drag. However a rocket can operate high above the earth's atmosphere, so in that case only the gravitation force must be considered. The strength of this force depends on how close the rocket is to a massive body such as the earth, the moon, or the sun. If the rocket is not close to any massive body (say, it is halfway between the earth and Mars), then the gravitational force acting on it is extremely small, less than one percent of what it is at the earth's surface, and consequently the rocket will continue to move at almost constant velocity without needing any thrust.

Newton's Second Law of Motion states that a body of mass M cannot accelerate unless acted upon by an unbalanced force F. When such a force is acting, the acceleration is directly proportional to the magnitude of the force F and inversely proportional to the mass M.

To fully understand this profound law, it is necessary to have a clear understanding of acceleration, force, and mass. Consider an automobile that can increase its speed from 20 miles per hour to 60 miles per hour in 10 seconds. In that case the acceleration is 40 miles per hour per ten seconds, or 4 miles per hour per second. Since there are 3600 seconds in an hour, we may convert this to 4/3600 or 1/900 mile per second per second. This should be perfectly clear, unless you question what is meant by distance and what is meant by time. Einstein showed that there is strange behavior of distance and time for objects moving near the velocity of light. However neither jet engines nor rockets can propel anything to such high velocities, so we can ignore the theory of relativity in this book.

What is force? We know about the force that can be exerted by our muscles. We know that a coiled spring exerts a force. A gas under high pressure exerts a force on its containing walls. When water in a pipe freezes, the ice expands and exerts force on the pipe, perhaps breaking it. When water issues from a large fire hose at the rate of 200 gallons per minute, the thrust force pushing the nozzle backward is about 100 pounds, and it takes two men to hold the hose. The sun exerts a gravitational force on the earth, and the earth exerts a gravitational force on a falling apple. So, we have a pretty good idea about what force is, but we need to have a quantitative way to specify it. For our first try, we say that a force of one pound is defined as the gravitational force acting on a mass of one pound (that is, the weight of the mass) on the earth's surface at sea level. However the gravitational force is different in different places; on the surface of the moon, the same pound of mass will only experience a gravitational force of 0.166 pounds so its weight would be less. Hence we would prefer a definition of force which could be verified anywhere in the universe by comparing a force with a standard.

Another way to appreciate the difference between mass and weight is to consider an astronaut weighing 170 pounds at sea level on earth, who is in orbit and engaged in extravehicular activity, while floating in space. His mass is still 170 pounds, but his weight is clearly zero.

For our next try to define force we use Newton's Second Law. We define the force of one pound as the force that will accelerate a mass of one pound by exactly 32.17405 feet per second per second. This can be used anywhere, as long as we also know the definitions of mass, length, and time. The value 32.17405 feet per second per second, called g_c, was selected because it is equal to the acceleration of gravity at sea level.

Length of a meter can be defined as the distance that light travels in vacuum during a time interval of 1/229,792,458 of a second. Time can be defined in terms of the frequency of a certain spectral line emitted by a cesium-133 atom. It has a frequency of 9,192,631,770 times a second. These are believed to be independent of location in the universe.

The traditional way of defining a standard for mass for many years was to compare unknown masses with the mass of a piece of platinum that was defined as being one kilogram or one pound. This standard is useless unless one can travel to Paris or Washington where the standard is maintained. Not so easy if one is on Mars. It would be possible to define a mass standard as the mass of a certain number of atoms of a specified type. Then, the unit of mass on Mars would be the same as on earth.

The Second Law is essential in understanding the acceleration of a rocket. The thrust force generated by the rocket exhaust acts to accelerate the mass of the rocket, which is changing as the rocket fuel is being consumed.

The Third Law of Motion states that for every action there is an equal and opposite reaction along the same straight line. Let us take some examples. A book lies on the table. The book has mass, and gravitation exerts a downward force on the book. According to this law, the table exerts an equal upward force on the book, and nothing moves. In a more complicated example, a bullet is fired from a gun. Let M be the mass of the gun, and m be the mass of the bullet. Let t be the time that the bullet is accelerating in the gun barrel. Let v_1 be the velocity at which the bullet leaves the gun, and let v_2 be the velocity of the gun after it recoils, assuming it is not restrained. According to the Second Law, the force acting on the bullet while it is accelerating is equal to mv_1/g_ct. The force acting on the gun, causing it to recoil, is Mv_2/g_ct. According to the Third Law, these two opposite forces must be equal to one another. Equating them, we obtain the relationship $v_1/v_2 = M/m$.

The dimensional constant g_c is equal to one and may be ignored if force is expressed in newtons, mass in kilograms, and velocity in meters per second. However, if thrust is expressed in pounds of force, mass flow rate in pounds of mass per second, and velocity in feet per second, then the dimensional constant is equal to 32.17.

How does all this apply to a moving fluid? The thrust force on the gun was mv_1/g_ct. For a moving fluid, m/t is the rate of mass flowing across a selected plane. Accordingly, the thrust caused by the moving fluid is v_1 times (m/t) divided by g_c.

For a rocket, the bullet is analogous to the stream of exhaust gases at velocity v_e, and the thrust force on the rocket itself is analogous to the recoil force on the gun. Thus, we have the basic equation for thrust of a rocket: thrust = v_e times (m/t) divided by g_c.

For a jet engine in level flight at a constant speed, things are a little different. The air entering the engine at the flight velocity v_f produces a force slowing down the plane, while the larger velocity of the exiting gas v_e produces an accelerating force. The net force overcomes the aerodynamic drag. The equation expressing this is:

Thrust force = drag force={[(m/t) times v_e] minus [(m minus m_{fuel})/t] times v_f} /g_c.

The additional term involving the rate of fuel flow is needed because the exit gases, containing fuel combustion products, have greater mass than the inlet air.

Newton's Law of Gravitation states that every mass attracts every other mass with a force that is directly proportional to the product of the two masses and inversely proportional to the square of the distance between them. Unless one of the two masses is very large, such as a planet, the force is extremely small. The distance from a planet to an exterior body is measured from the center of the planet.

For a jet plane, this law tells us how much lift the wings must generate to balance the downward gravitational force. For a rocket, this law is needed to calculate what velocity must be attained to go into earth orbit, and what velocity is needed to escape the earth, the moon, or any other body. (See Appendix B.) For a ballistic rocket aimed at a target on earth, the law of gravitation along with the laws of motion is needed to calculate the trajectory.

APPENDIX B

THE MATHEMATICS OF ROCKETRY AND SPACE TRAVEL

The formulas governing earth orbits and escape from the earth follow directly from Sir Isaac Newton's formulations of the laws of gravitation and motion. These were worked out in the late seventeenth century. Hence, the knowledge of what is required to send a spacecraft into orbit or to the moon has been available for a long time, but could not be used until powerful rockets were available (mid-twentieth century).

For a spacecraft orbiting the earth, the necessary velocity is determined by a balance between the gravitational force trying to pull the object to earth and the centrifugal force trying to make it fly off into space. The gravitational force is given by Newton's formula:

$$g_e(r_e^2/r_o^2)M/g_c. \qquad (1)$$

In this formula, g_e is the acceleration due to gravity at the earth's surface, which equals 32.2 feet per second per second in English units or 9.8 meters per second per second in S.I. units. The symbol r_o is the distance from the center of the earth to the orbiting spacecraft, r_e is the distance from the center of the earth to the surface of the earth, M is the mass of the spacecraft, in pounds or kilograms, and g_c is a constant equal to unity in S.I. units or 32.2 in English units.

To be in orbit, this force is balanced by the centrifugal force:

$$Mv^2/r_o g_c.$$

Here, v_o is the velocity of the spacecraft around its orbit in feet per second or meters per second. When the centrifugal force and the gravitational force are equated, the mass of the spacecraft M and the constant g_c are seen to cancel out. The resulting equation may be rearranged to:

$$v_o = r_e(g_e/r_o)^{1/2} \qquad (2)$$

The orbital velocity is inversely proportional to the square root of the distance from the spacecraft to the center of the earth. The spacecraft must be high enough so that atmospheric resistance is small; it is quite small at 100 miles altitude and negligible at 200 miles altitude. To calculate the orbital velocity at 200 miles (1,056,000 feet) altitude, multiply 20,890,000 by the square root of [32.2 divided by (1,056 ,000 plus 20,890,000)] , obtaining 25,300 feet per second (4.79 miles per second). (The number 20,890,000 is the mean radius of the earth in feet.)

Next, let us calculate the velocity needed to escape the earth's gravitational field entirely, which is referred to as the escape velocity. The easiest way to do this is to imagine a mass M, initially at rest extremely far from earth, to be released so that it can fall to earth. In this calculation, air resistance is ignored. (To consider air resistance, we would have to know the aerodynamic drag on the object.) The object will accelerate as it falls, with an increase in kinetic energy. If v_e is its velocity when it reaches earth, its kinetic energy just before arriving is:

$$Mv_e^2/2g_c$$

according to Newton's laws of motion. This kinetic energy is also the difference in potential energy of the mass M at the earth's surface, caused by the earth's gravitational field, and the same mass very far away, where the earth's gravitational field has zero potential energy. This potential energy difference is exactly the mechanical work that has to be done to move mass M from the earth's surface to very far away (neglecting air resistance) and may be expressed as:

the integral of Fdx, from the earth's surface to infinity, where F is the gravitational force of Equation (1).

When we equate this integral to the kinetic energy of a falling mass ($Mv_e^2/2g_c$), and integrate, and, finally, solve for v_e, we obtain

$$v_e = (2g_e r_e)^{1/2} \qquad (3)$$

where v_e is the escape velocity and g_e is the acceleration due to gravity at the earth's surface. By comparing equations (2) and (3), we see that the escape velocity is closely related to the orbital velocity. They differ by only the square root of the ratio $2r_e/r_o$.

Introduction of numerical values into equation (3) yields an escape velocity of 36,670 feet per second or 6.95 miles per second.

Note that the only data needed to make this calculation were the acceleration of gravity at the earth's surface and the radius of the earth.

Note also that if a rocket were fired eastward near the equator, it would get an advantage of about 0.3 miles per second because of the earth's rotation.

Now we come to a different question. What is required to accelerate a rocket to a high velocity? This depends primarily on three factors: the ratio of launch weight to burnout weight, the nature of the propellant or propellant combination, and the combustion chamber pressure. There is also a lesser effect of the outside pressure and whether the diverging rocket nozzle diverges sufficiently for the exhaust pressure to be equal to the outside pressure. Let us look at the basic equation of rocket propulsion.

Let M_i and M_f denote the initial and final masses of the rocket, the difference being the propellant consumed. Let v be the velocity of the rocket at any instant, relative to its initial velocity, if any. Let v_x be the velocity at which the combustion products leave the nozzle (i.e., the exhaust velocity). Then, from Newton's laws of motion, the thrust force exerted on the rocket is given by:

$$v_x(dM/dt)/g_c,$$

where dM/dt is the rate of propellant flow. According to Newton, the acceleration of the rocket is equal to the thrust force divided by the mass at any instant M/g_c. (This assumes no air resistance.) The acceleration is dv/dt. Then we have

$$dv/dt = v_x(dM/dt)/M.$$

This differential equation may be integrated between the limits of M_f and M_i for M, and v_i and v_f for v. The result, when rearranged, is:

$$v_f - v_i = v_x \ln(M_f/M_i). \quad (4)$$

Equation (4) is very important in rocketry; it tells us that the increase in velocity achievable by a rocket depends on two things: the exhaust velocity v_x and the mass ratio M_f/Mi. The mass ratio is controlled by the designer of the rocket. The exhaust velocity is approximately equal to the specific impulse of the propellant multiplied by g_e.

What is specific impulse? It may be defined as the thrust produced at a given instant divided by the mass rate of propellant consumption at that instant. It has the dimension of seconds, in either English or S.I. units. It depends not only on the propellant, but also on the pressure in the combustion chamber, the outside pressure, and the nozzle design. In order to be able to compare various propellants on an equal basis, the specific impulse may be measured or calculated at "standard conditions", namely 1000 psi chamber pressure and outside pressure of one atmosphere. It is possible to calculate how the specific impulse will change at other conditions.

The fact that the specific impulse at standard conditions of any propellant may be calculated only from knowledge of the chemical constituents of the propellant is surprising, and very useful in the process of developing a propellant. For example, to find the optimum fuel-oxidant ratio, there is no necessity of doing tests; instead, calculations give the answer. The calculations are quite complex and require the use of a computer. They are described later.

A simple formula may be developed for estimating the specific impulse of any propellant if we can estimate the mean molecular weight of the exhaust gases and the temperature in the combustion chamber. Let M be the mass in kilograms of one mole of combustion products (i.e., the mean molecular weight divided by 1000); let T be the combustion chamber temperature, in Kelvins; and let R be the universal gas constant (8.33, if mass is in kilograms, velocity is in meters per second, and T is in Kelvins). These are S.I. units. Now, we make several simplifying assumptions. First, we assume that the rocket nozzle is able to convert 70 percent of the thermal energy of the gases in the combustion chamber to kinetic energy. Second, we assume that the thermal energy of the hot combustion gases is 3RT per mole. Now, we equate the thermal energy per mole to the kinetic energy per mole:

$$0.7(3RT) = M\, v_x{}^{1}/{}^{2}/2.$$

Rearranging this equation gives: $v_x = (4.2RT/M)^{1/2}$. (5)

To use this equation, we must assume a representative value of combustion chamber temperature, which we take to be 2200 K, and a representative value of M, taken to be 0.00667 kilograms per mole. Then, we can calculate that the exhaust velocity is 3400 meters per second, or 11,150 feet per second, or 2.1 miles per second.

If we want to know the corresponding specific impulse, we may use English units and divide 11,150 by 32.2, obtaining 346 seconds. Or, we may use metric units, dividing 3400 by 9.8, again obtaining 346 seconds. The specific impulse is directly proportional to the exhaust velocity.

If we take a look at equation (5), we see that the exhaust velocity is proportional to the square root of the ratio of the combustion chamber temperature to the mean molecular weight of the exhaust gases. It is interesting to note that the velocity of gases at the throat of the nozzle is the velocity of sound, which also varies with the square root of the ratio of temperature to molecular weight.

Equation (5) gives us information about the highest specific impulse obtainable from any chemical propellant system. The flame temperature will never be much over 3000 K, because above this temperature the product molecules will dissociate appreciably, absorbing energy. The mean molecular weight could not be below 2, which is the molecular weight of hydrogen. Obviously it must always be appreciably above 2, because the hydrogen must react with something (oxygen or perhaps fluorine) in order to generate heat. Then, water vapor (molecular weight 18) or hydrogen fluoride (molecular weight 20) must be a prominent constituent of the exhaust gases. If an excess of hydrogen is provided, this lowers the mean molecular weight but also lowers the combustion temperature.

One may conclude that the search for a chemical superfuel very much better than hydrogen-oxygen has no chance of success.

While equation (5) is very useful in estimating what sort of rocket performance one might expect from some initial combination of chemicals, it is not at all precise. Its use requires an estimate (or a guess) as to the temperature in the combustion chamber, as well as an estimate of the mean molecular weight. There is a very precise way of calculating the theoretical exhaust velocity by use of a digital computer. The actual exhaust velocity will always be a little less than the theoretical exhaust velocity.

This accurate calculation involves two steps; the determination of the temperature and composition of the hot gases in the combustion chamber, and the expansion process in the nozzle. It might be assumed that the hot gases in the combustion chamber will be in a state of chemical equilibrium, and that no further chemical changes occur during the nozzle expansion. There is no friction, no heat losses, and the direction of the exit velocity is precisely in line with the principal axis of the rocket. Any condensed-phase particles in the combustion products (possibly aluminum oxide) maintain the same velocity as the gases during the expansion.

After such a calculation is carried out, further refined calculations may be made regarding each of the assumptions listed. Without such refinements, the resulting exhaust velocity will be perhaps ten percent below the ideal velocity, and, with all possible refinements, the result will be within one percent of the ideal velocity.

The calculation of the equilibrium in the combustion chamber generally will give results very close to the actual behavior, because of the very high temperature causing the chemical reactions to come very close to the equilibrium values. This calculation involves the following elements:

The law of conservation of species applies. That is, the ratio of hydrogen atoms to oxygen atoms, for example, in the equilibrium products is exactly the same as in the reactants.

The principle of conservation of energy applies. That is, the energy associated with each species in the products, when summed, is equal to the potential energy of the species in the reactants.

It is assumed that we know what chemical species are present in the combustion chamber, including dissociated species such as OH, H, O, and AlO, and that they are in chemical equilibrium with one another. The composition is determined by the requirement that the free energy (known for each species) is a minimum at the prevailing temperature.

The temperature must be consistent with the conservation of energy, the concentration of each species, and the specific heat of each species (which depends on temperature).

A computer formulates all these elements in the form of equations, consults tables of free energy and specific heat for all species, and then finds the unique solution that satisfies all the above restraints.

The next step is to calculate what happens to this hot mixture of gases (and possibly liquid droplets) during the nozzle expansion. If it is assumed that no further chemical changes occur during the expansion (frozen equilibrium), the calculation is relatively simple. The expansion is assumed to be isentropic, and the perfect gas law is obeyed. The result is the exhaust velocity.

One way to refine the calculation is to assume that chemical reactions occur during the expansion, maintaining a shifting equilibrium because of the decreasing temperature and pressure as the gas expands. The particles, if present, are assumed to keep up with the gas as it accelerates. This is known as the shifting equilibrium treatment, and the result gives a value of exhaust velocity somewhat higher than the frozen equilibrium result previously described.

The truth is somewhere in between the frozen and shifting equilibrium treatments, because the expansion is so rapid that the chemical changes may not be able to keep up. The next level of refinement is to consider the chemical kinetics of the recombination reactions, and also to consider the drag forces on the entrained particles (which depend on their size!).

Here is a sample of calculation results that assumed shifting equilibrium for a hydrogen-oxygen rocket with an oxygen-hydrogen mass ratio of 5.55. The exhaust pressure was assumed to be one percent of the combustion chamber pressure.

Combustion chamber temperature	3346 K
Exhaust temperature	1786 K
Exhaust velocity	12,652 feet per second (2.40 miles/sec.)
Specific impulse	394 seconds
Exhaust area/throat area	3.29

The composition of the gases (mole fractions) is shown for the combustion chamber and the exhaust:

	Combustion chamber	Exhaust
Water vapor	0.6386	0.6993
Molecular hydrogen	0.2938	0.3004
Atomic hydrogen	0.0334	0.0002
Hydroxyl radical	0.0304	0.0000
Molecular oxygen	0.0017	0.0000
Atomic oxygen	0.0020	0.0000

The calculation also showed very tiny amounts of hydrogen peroxide and the radical HO_2.

When a similar calculation is made for a solid propellant containing ammonium perchlorate, aluminum, and a hydrocarbon polymer binder, the calculation is more complex, because six chemical elements are involved instead of two, and also both liquid and solid aluminum oxide occur. The calculation was done for a combustion chamber pressure of 1000 psi and an exhaust pressure of one atmosphere. The combustion chamber temperature came out to be 3323 K and the exhaust temperature 2136 K. The specific impulse came out to be 256 seconds. Over one hundred species were considered in the calculation, but all but twenty of them showed up in very tiny quantities in the combustion chamber and are not shown in the list below, showing mole fractions in the combustion chamber. (Note that aluminum oxide is liquid in the combustion chamber but solid in the exhaust. The heat of solidification was taken into account.)

Molecular hydrogen	0.3238
Carbon monoxide	0.2237
Hydrogen chloride	0.1190
Nitrogen	0.09886
Aluminum oxide (liquid)	0.09378
Water vapor	0.08960
Atomic hydrogen	0.02525
Carbon dioxide	0.00790
Atomic chlorine	0.00620
Aluminum monochloride	0.00499
Hydroxyl radical	0.00297
Aluminum dichloride	0.00167
Aluminum oxychloride AlOCl	0.00095
Aluminum hydroxide AlOH	0.00032
Aluminum trichloride	0.00023
Nitric oxide	0.00021
Atomic oxygen	0.00014
Aluminum vapor	0.00009

Aluminum monoxide	0.00009
Aluminum dioxide	0.00009

Once the computer program has been set up, it is easy to repeat such calculations for different chamber pressures or different exhaust pressures, as well as for different proportions of the propellant ingredients.

APPENDIX C

SELECTED BIOGRAPHIES OF MAJOR CONTRIBUTORS TO JET PROPULSION AND ROCKETRY

ISAAC NEWTON

Until Newton formulated his equations, there was no way which anyone could calculate the velocity needed to escape the earth or the velocity a rocket could reach under ideal conditions. After his equations were published, such calculations could be easily made.

Newton was born in 1642, at Woolsthorpe, Lincolnshire, England. His father, who died before he was born, was a farmer, and probably illiterate. His mother remarried, and Newton was shipped off to some relatives to be raised. After this inauspicious beginning, no one could imagine that Newton would become possibly the greatest scientist who ever lived.

He obtained a scholarship and entered Cambridge University. He graduated with a bachelor of science degree in 1665. He then left the university and pondered the laws of nature, in solitude. The key problem he was concerned with was the orbits of the planets.

Previously, Copernicus had discovered that the earth and other planets revolved around the sun instead of, as previously thought, the earth being the center of the universe and everything revolving around it. Then, Kepler followed up this work with the discovery that the planets moved in ellipses rather than circles, the sun being at one of the two foci of each ellipse. Further, he discovered that the time to complete an orbit about the sun was proportional to the three-halves power of the mean distance of a planet from the sun. No one had any explanation of these findings.

In order to solve the problem, Newton had to invent the calculus, which he did. Then, he had to formulate the laws of motion and the law of universal gravitation. He could then derive Kepler's findings. Amazingly, after making these discoveries he did not publish them for twenty years.

He returned to Cambridge in 1667, and was appointed Lucasian professor of mathematics in 1669. Obviously, his colleagues recognized his genius.

He went on to show that falling bodies on earth followed the same laws as the planets. He provided an explanation of the tides. He calculated that the earth was flatter at the poles than at the equator. Finally, in 1687 he published "Mathematical Principles of Natural Philosophy", which is generally agreed to be the most important book in the history of science.

He also made revolutionary discoveries on the nature of light. He observed how a prism refracted different colors of light differently. This provided an explanation of rainbows. More important to astronomers was his invention of the first reflecting telescope. Previous telescopes required lenses to focus light; the various colors were focused differently after passing through a lens (chromatic aberration), resulting in fuzzy images. Newton found that he could use a concave mirror instead of a lens and avoid this problem. He published a theory of the nature of light, and wrote a book on this subject.

He became widely recognized throughout Europe. His work showed that the mysteries of nature could be explained rationally. He was elected President of the Royal Society in 1703. He was the first British scientist to be knighted. He died in 1727, and is buried in Westminster Abbey.

GEORGE CAYLEY

Sir George Cayley, born in Yorkshire, England in 1773, led the way to development of heavier-than-air manned gliders. He was active in a variety of engineering projects. His aeronautical experiments led to design of an efficient cambered (curved) airfoil, and included many tests with model gliders. He concluded that a tail assembly was needed. Some time prior to 1849 he designed a triplane which carried a ten-year old boy in a successful flight, but no details are known. In 1853 he built a larger glider, in which one of his employees flew.

His work stimulated others to experiment with gliders. He died in 1857.

OTTO LILIENTHAL

Lilienthal, born in Germany in 1848, was the first person to make repeated short glider flights in a fixed-wing heavier-than-air glider. He was called the "Glider King", having made over 2000 flights in gliders of his own design. Some of his flights were photographed. His gliders were controlled by body shifting. (The Wright brothers later concluded that this method of control was insufficient.) He died in a glider crash in 1896. The Wright brothers credited him as a major inspiration to them.

ORVILLE AND WILBUR WRIGHT

Wilbur and Orville Wright were born in the Midwest United States in 1867 and 1871 respectively. They both attended high school but had no higher education. In 1892 they opened a bicycle repair and sales shop in Ohio. By 1896 they began manufacturing their own brand of bicycle.

They became deeply interested in aviation in 1899. They read what was available and concluded that manned, powered flight required three things, an engine with propeller, wings, and means of control. They felt that the really difficult one was means of control. They planned a series of glider flights until this problem was solved to their satisfaction, after which they would add an engine.

Their first manned glider was tested at Kitty Hawk, North Carolina in 1900 with unsatisfactory results. Their second glider, with larger wings, was tested the following year, but was not controlled adequately;

it was unable to make turns. They were using a technique of warping the wings to permit banked turns, but this did not work very well. At this point, they built a wind tunnel in their bicycle shop, and tested numerous models of wings. They discovered the benefit of longer, narrower wings; that is to say, a larger aspect ratio. They had a forward elevator. They added a rear, vertical, moveable rudder. Their next model incorporated these features, and on October 8, 1902 they made their first flight with true control. They made over 700 glides, the longest covering 622 feet. In March 2003 they applied for a patent on their method of controlling an airplane.

The next step was to add an engine and a propeller. Their mechanic, Charlie Taylor, built an aluminum gasoline engine in their bicycle shop. They carved their own wooden propeller. They incorporated these elements in a plane, the Wright Flyer, which they took to Kitty Hawk and made history. They made a series of four flights, the longest being 852 feet, in December 1903. These flights were photographed and witnessed.

They returned to Ohio, built another Flyer, and made 105 flights in 1904. On September 20, 1904, Wilbur made a flight lasting 39 minutes, including flying the plane in a circle. At this point they considered their plane to be proven.

In 1908 they signed contracts with the United States Army and a French company, and then made a series of demonstration flights in France which gave them world recognition. On return to the United States, they were invited to the White House by President Taft, where they received awards. Wilbur died in 1912, while Orville lived on until 1948.

GLENN H. CURTISS

Glenn Curtiss was born on May 21, 1878, in Hammondsport, New York. He started a bicycle shop at about the same time as the Wrights. He developed an engine for a bicycle, making it a motorcycle, and set a world's record of 64 miles per hour with it in 1903. He decided to manufacture airplane engines, starting the Curtiss Manufacturing Company in 1905. He met the Wright brothers in 1906. He and the Wrights had some patent disputes that were finally settled, resulting in the Curtiss-Wright Corporation, which still exists. In 1909 he won a competition at the world's first air meet in Rheims, France, flying 6 miles at 46.5 miles per hour. In 1910 he worked with the U.S. Navy to build the first plane to take off from a ship. In 1911 he built the first seaplane, which had both wheels and floats. He designed planes for the U.S. Army during World War I. He was recognized as the father of the U.S. aviation industry. He died on July 23, 1930.

FRANK WHITTLE

Frank Whittle was born near Coventry, England in 1907. At the age of 16, he joined the Royal Air Force as an apprentice. His abilities were recognized and he rose rapidly in the RAF, joining a fighter squadron in 1928 and becoming a test pilot in 1931.

In 1929 he conceived the idea of a turbojet engine. He received a patent in 1930. He had difficulty finding support to build such an engine, but he finally built an engine with a single-stage centrifugal compressor and a single-stage turbine. This was a laboratory test rig which demonstrated the feasibility of the concept in April 1937. He became associated with the firm Power Jets Ltd., which received a government contract in July 1939 to build such an engine for a small experimental aircraft. This plane flew for the first time in 1941.

However, Dr. von Ohain in Germany developed a turbojet engine which first flew two years earlier.

HANS VON OHAIN

Hans von Ohain, born in 1911, obtained a doctorate in physics at the University of Gottingen, Germany. Ernest Heinkel, the German aircraft builder, learned that von Ohain had conceived the idea of a turbojet engine in 1933, so he hired him. Von Ohain patented his design in 1934. A successful bench test of his engine was accomplished in September 1937 (five months after Whittle). It is believed that von Ohain was not aware of Whittle's work. The Heinkel HE 178 was built to serve as a test bed for von Ohain's engine. This plane first flew on August 27, 1939 by Captain Erich Warsitz, and this was clearly the first flight of a plane with a turbojet engine. Improved turbojet planes were developed in Germany, and in 1944 the Messerschmitt ME262 turbojet fighter plane was put into service in World War II.

CHUCK YEAGER

Major General Charles Elwood Yeager was born in February 1923 in West Virginia. He joined the U.S. Army Air Force as a private in 1941. He became a fighter pilot, and shot down 11.5 enemy planes, including a Messerschmitt ME 262 jet plane. After the war he became an Army test pilot.

He was selected to fly the Bell X-1 rocket plane through the sound barrier, becoming the first person flying a plane in level flight at a speed greater than Mach 1, on October 14, 1947. Then, on December 12, 1953, he, together with Jack Ridley, achieved the speed of Mach 2.44.

In 1969 he was promoted to Brigadier General and was assigned as vice commander of the Seventeenth Air Force.

KONSTANTIN TSIOLKOVSKY

Konstantin Eduardovich Tsiolkovsky was born in Russia in 1857. He was the son of a Polish deportee to Siberia. He was active in aeronautics and astronautics throughout his career, as a teacher and researcher. As early as 1894 he designed an airplane (thirteen years before the Wright brothers), which flew in 1915. He was the first in Russia to build a wind tunnel (1897).

He was interested in rocketry from an early age. In 1903 he published an article in a Russian aviation magazine which contained the formula he derived showing how the velocity achieved by a rocket depended on the ratio of mass of a rocket before and after its fuel was consumed. He also showed how a multiple-stage rocket could reach very high velocities and could escape from the earth, while a single-stage rocket could never do this. He was aware that liquid propellants were more effective than solid propellants, and demonstrated this. He was the author of "Investigations of Outer Space by Rocket Devices" (1911) and "Aims of Astronauts" (1914). He died in 1935.

Here are quotes from his writings: "Mankind will not forever remain on earth, but in the pursuit of light and space will first timidly emerge from the bounds of the atmosphere, and then advance until he has conquered the whole of circumsolar space." Another one is: "Earth is the cradle of humanity, but one cannot remain in the cradle forever."

Chronologically, he was well ahead of Goddard and Oberth (to be reviewed later), and it is not clear to what degree they were aware of his work. Tsiolkovsky is sometimes referred to as the theoretical father of rocketry

GUSTAF DE LAVAL

By the mid-nineteenth century it was known that electricity could be generated by rotating a series of conductors through a magnetic field. However, force is needed to turn the generator. The steam engine with pistons was available as a source of power, but it was suspected that a steam turbine would be a more efficient way of powering the generator. Gustaf de Laval, a Swedish engineer born in 1845, was successful in designing an efficient steam turbine, still in use today. A key element of this invention is the nozzle, which converts the initial high pressure of the steam to high velocity. This de Laval nozzle allows the high-pressure steam to flow into a converging nozzle. The steam decreases in pressure and temperature and increases in velocity until it reaches the throat of the nozzle. At this point the steam is at about half the initial pressure and is moving at the velocity of sound. De Laval found that if one adds a divergent section to the nozzle at the throat, the steam will continue to decrease in pressure and increase in velocity, reaching a speed of several times the velocity of sound. This high-velocity jet impinges on a turbine blade and provides the motive power.

It turns out that this kind of nozzle is exactly what is needed in a high-performance rocket, which Robert Goddard (below) recognized and incorporated in his rockets. De Laval died in 1913, apparently without thinking about rockets.

ROBERT HUTCHINS GODDARD

Goddard was born in Worcester, Massachusetts in 1882. He became interested in space flight at an early age. After undergraduate studies at Worcester Polytechnic Institute, he obtained a master's degree in physics at Clark University in Worcester and joined the faculty there. He obtained two United States patents in 1914 (#1,102,653 and #1,103,503), which dealt with multiple-stage rockets, liquid propellants, and expansion nozzles. He was aware of de Laval's work with nozzles (see above). At this time no one had successfully launched a liquid-propellant rocket, and Goddard was determined to do so. In 1920, he published "A Method of Reaching Extreme Altitudes", which came to the attention of Hermann Oberth (see below).

In 1917 Goddard obtained a grant of $5000 from the Smithsonian Institution in Washington. He experimented with rockets for the next nine years, and launched his first liquid-fuel rocket in March 1926. The test was unsuccessful, only lasting less than three seconds before the rocket crashed.

However, without much financial or technical support, he persevered, and attracted the interest of Charles Lindbergh. Lindbergh helped him find financial support, and his work continued, with many attempts to launch rockets. The most successful flight reached 9000 feet. During this time, Goddard was granted over 200 patents. He wrote an article suggesting that rockets could reach the moon. The New York Times editorial page ridiculed this, stating that rockets could not work in space because they would have nothing to push against. When NASA sent rockets to the moon in 1969, the New York Times published an apology for its mistake.

Goddard, the first person to launch a liquid-propellant rocket, died in 1945.

HERMANN OBERTH

Hermann Julius Oberth was born in 1894 in what is now Romania. At the age of eleven, he became fascinated by the concept of space travel after reading Jules Verne's "From the Earth to the Moon". He went to Germany to study medicine, but remained fascinated by space. He changed his field of study to physics. In 1922, he submitted a doctoral dissertation on rocket science, which was rejected as "utopian". Nevertheless, he published it as a book, "By Rocket into Interplanetary Space". A few years later, he was finally awarded a doctoral degree at Babes-Bolyai University, in Romania. He then joined the faculty of the Technical University of Berlin, Germany.

In 1928 and 1929 he was scientific consultant on the first film to have scenes in space, "Frau im Mond" (The Woman in the Moon), directed by Fritz Lang. He became a member of the Verein fur Raumschiffahrt (Spaceflight Society) and acted as a mentor to rocket enthusiasts. In 1929, helped by his students at the university, he launched his first liquid—propellant rocket. One of his students was Werner von Braun (see below). Oberth is known to have been stimulated by Goddard's writings (see above).

He continued his rocket experiments in the 1930s. World War II approached, and he worked on the V2 rocket with von Braun at Peenemuende. After a while, he left this project to develop solid-propellant anti-aircraft rockets at Wittenberg. After the war, he went to Italy, where he continued his work on solid-propellant rockets. In 1953 he published his book, "Man in Space". Soon thereafter, he joined von Braun in Huntsville, Alabama to work on space rockets. Later, he worked in the United States for Convair as a consultant on the Atlas rocket.

He died in 1989 at the age of 95. He is recognized as one of the founding fathers of rocketry and astronautics.

WERNER VON BRAUN

He was born in Wirsitz, Germany in 1912. He obtained a B.S. degree in mechanical engineering in 1932 and a doctorate in physics in 1934, at the age of 22, in each case from the University of Berlin. His first involvement in rockets occurred in 1930, when, at the age of eighteen, he assisted Hermann Oberth in experiments with liquid-propellant rockets. Then he was instrumental in setting up a small development station for liquid fuel rockets sponsored by the German Society for Space Travel. He worked at this station in his spare time until 1932. At this point the German Ordnance Department showed interest in von Braun's work and supported his setting up a small rocket development program at Kummersdorf Army Proving Ground, with one mechanic. In 1934 he successfully launched two liquid fuel rockets which reached altitudes of 1.6 miles. By 1937, his group had grown to 80 people.

In 1937 he was made Technical Director of a new Liquid Fuel Rocket and Guided Missile Center at Peenemunde, where he remained until the end of World War II, in 1945. At its peak of activity, 10,000 people worked at this center. By 1940 they had conducted 25 successful launchings of A5 rockets, which reached 10 miles altitude and had a range of 12 miles. They then moved into the development of the V2 rocket, which had a much greater range. The first successful launching was in 1942. This led to the manufacture of two thousand V2 rockets, about five hundred of which landed in London, causing considerable damage. From 1943 to 1945, an anti-aircraft guided missile was developed by the Peenemuende group. There were 44 successful launchings.

The war ended, and von Braun, considered the top expert in rocketry, was invited to the United States, where he was made Project Director of a U.S. Army program to develop rockets, at Fort Bliss, Texas. A number of his German associates were brought to the U.S. and joined the group. In 1950 this entire program was transferred to Redstone Arsenal, Huntsville, Alabama. Von Braun was Technical Director of the Guided Missile Development Group. In 1952 he was made Chief, Guided Missile Development Division, Redstone Arsenal. His group developed a series of rockets, leading to NASA's Apollo Program. He became a U.S. citizen. Under NASA auspices he became director of the Marshall Space Flight Center, Huntsville, Alabama in 1960.

In 1972 he left NASA to become vice president of Fairchild Industries in Maryland. He retired in 1977, at the age of 65, and died shortly thereafter.

Von Braun, on the foundation supplied by Tsiolkovsky, Goddard, and Oberth, proved conclusively that reliable liquid-propellant rockets could be made.

SERGEI P. KOROLEV

While the Germans were experimenting with liquid-propellant rockets in the 1930's, some Russians were also active in this field, under the leadership of Sergei Korolev. Born in1906, he is considered to be the founder of the Soviet space program. He was educated in aeronautical engineering at Kiev Polytechnic Institute. In 1930 he co-founded a "Group for Investigation of Reactive Motion" in Moscow. They tested liquid-fuel rockets. He wrote a book, "Rocket Flight in the Stratosphere".

After two years, the Soviet government took over this group and replaced it with a "Reaction Propulsion Scientific Research Group", in which Korolev was a key figure. They developed rocket-propelled missiles and gliders. They also developed RP-318, a rocket-propelled aircraft. While the group was strongly interested in liquid propellants, potentially needed for space travel, the Soviet government wanted them to concentrate on solid propellants, which were more useful for military weapons.

As a result of this dispute, Korolev and some of his associates were imprisoned in Siberia during the war. However, after a few years, he and other aeronautical specialists in prison were given permission to work with rockets to develop weapons. This continued through the war. After the war, Korolev was released from prison and put in charge of developing a long-range ballistic missile. Undoubtedly the Russians were responding to the German V2 development.

After the war, Korolov and others were engaged in a secret program to develop an intercontinental ballistic missile capable of carrying a nuclear warhead. Korolev led the way to the design of the ICBM that launched the first Sputnik. He developed the Vostok spaceship which carried Yuri Gagarin into orbit.He then led a program attempting to achieve a moon landing before the Americans. He was awarded the Order of Lenin, and was elected to the Russian Academy of Sciences. He died after a botched operation in 1966, and was buried in the Kremlin wall. Statues honoring him are to be found in Russia.

ARTHUR C. CLARKE

Clarke was born in Somerset, England in 1917. After a secondary school education, he became a radar specialist during World War II. In 1945 he published a proposal, far ahead of his time, to launch geostationary satellites to serve as communication relays. However, it turns out that the basic idea for this may be found in Oberth's book written in 1923. Such satellites were actually launched in 1962.

Clarke was a member of the British Interplanetary Society since 1936, and subsequently its chairman.

Clarke became a novelist specializing in science fiction, publishing 33 novels. He is recognized as a top science fiction writer, along with Isaac Asimov and Robert Heinlein. His writings have undoubtedly caused many people to develop an interest in space travel. One of his novels provided a detailed description of a space elevator, described previously in this book.

Clarke's best-known work, "2001: A Space Odyssey" was developed jointly with Stanley Kubrick, starting in 1964. It was simultaneously issued as a motion picture and a novel. His other novels include "Childhood's End", "The City and the Stars", and "Rendezvous with Rama".

Clarke lived in Sri Lanka for many years, and died in 2008.

YURI ALEKSEYEVICH GAGARIN

Gagarin, the first person to orbit the earth, was born in 1934 in Klushino, Russia. His parents worked on a collective farm. His father was a carpenter. In 1955 he joined the Russian Air Force and entered a program of military flight training. He became a fighter pilot. He was selected as an astronaut in 1960, perhaps because he was only five feet two inches tall. There was limited space in the early launch vehicles.

On April 12, 1961 he made his epic flight, lasting 108 minutes. After re-entry, he parachuted to earth. He promptly became a world celebrity. He was made deputy training director at the cosmonaut facility (Star City). He died in 1968 at the age of 34 as the result of a MiG plane crash.

APPENDIX D

BIBLIOGRAPHY

1. Cumpsty, N. and Cumpsty, N.A., Jet Propulsion: A Simple Guide to the Aerodynamic and Thermodynamic Design and Performance of Jet Engines. 2003.
2. Flack, R.D., Fundamentals and Applications of Jet Propulsion. 2005.
3. Hesse, W.J. and Mumford, N.V.S., Jet Propulsion for Aerospace Applications. 1964.
4. Ley, W., Rockets, Missiles, and Space Travel. 1994.
5. Mattingly, J.D., Elements of Propulsion: Gas Turbines and Rockets. 2006.
6. Roy, G., Chemical Propulsion. 2005.
7. Sutton, G.P. and Biblarz, O., Rocket Propulsion Elements. Seventh Editiom. 2001.
8. Van Riper, A.B., Rockets and Missiles: The Life Story of a Technology.

Note: Any of the above books can be obtained online from Amazon or Barnes and Noble.

www.ingramcontent.com/pod-product-compliance
Lightning Source LLC
Chambersburg PA
CBHW022008170526
45157CB00003B/1196

* 9 7 8 1 4 5 0 0 6 5 8 9 4 *